Joseph D. Fehribach
Kirchhoff Graphs

Also of Interest

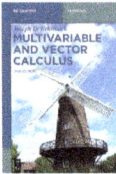

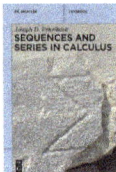

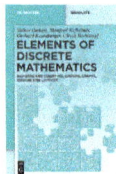

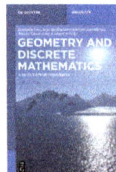

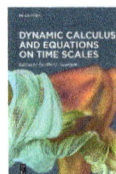

Joseph D. Fehribach

Kirchhoff Graphs

Vector Dependencies and Implications

DE GRUYTER

Mathematics Subject Classification 2020
05C20, 90B10

Author
Prof. Dr. Joseph D. Fehribach
Worcester Polytechnic Institute
Dept. of Mathematical Sciences
100 Institute Road
Worcester MA 01609-2280
United States of America
bach@math.wpi.edu

ISBN 978-3-11-140624-4
e-ISBN (PDF) 978-3-11-140857-6
e-ISBN (EPUB) 978-3-11-140939-9

Library of Congress Control Number: 2025943144

Bibliographic information published by the Deutsche Nationalbibliothek
The Deutsche Nationalbibliothek lists this publication in the Deutsche Nationalbibliografie;
detailed bibliographic data are available on the Internet at http://dnb.dnb.de.

www.degruyterbrill.com
Questions about General Product Safety Regulation:
productsafety@degruyterbrill.com

To my students and colleagues who have worked with me over the years on studying Kirchhoff graphs—particularly

- Ravindra Datta
- Ilie Fishtik
- Judi J. McDonald
- Tyler M. Reese
- Brigitte Servatius
- Randy C. Paffenroth
- Günter Rote
- Ronald L. Grimm
- Marcel Gietzmann-Sanders
- Jessica Kejia Wang
- Venkatraman Varatharajan
- Grace M. Baumgartner
- Evan G. Sayer

Gustav Kirchhoff (1824–1887).
(This work is in the public domain. https://upload.wikimedia.org/wikipedia/commons/f/fe/Gustav_Robert_Kirchhoff.jpg)

Foreword

Approximately thirty years ago, I was trying to understand an electrochemical reaction network associated with a molten carbonate fuelcell (MCFC). To do this, I drew a diagram for the network that my colleague Kas Hemmes found very interesting and that eventually helped lead to the patenting of a new fuelcell electrode. Since then, I was encouraged on several occasions to try to find ways to depict reaction networks more effectively. The concept of a Kirchhoff graph came out of these efforts. It was defined as a more-linear-algebraically exact variation on the reaction route graphs discussed by Illie Fishtik and Ravi Datta. My work on them began around 2005 and has continued in various forms since then.

At least in terms of their basics, there is nothing particularly deep mathematically in Kirchhoff graphs. They can be understood with just a strong undergraduate background in linear algebra and a basic familiarity with graph theory. Proving some results such as Kirchhoff graph uniformity does require a bit more sophistication, but much of what is discussed here could have been proven one hundred fifty years ago or so. Indeed, it would seem that Gustav Kirchhoff for whom these graphs are named could have proven many of these results. That he did not is likely due to the fact that a modern matrix and vector notation was not available in the mid 1880s when he finished his work. Modern notation and the use of matrices did not become standard until the early twentieth century, although some forms where used by Caley and Frobenius during Kirchhoff's lifetime.

What they lack in deep mathematics, however, Kirchhoff graphs make up for in their simplicity and beauty, at least in my view. They show how important it can be that the same edge vector can appear in multiple places in a graph.

The list of people that I owe thanks to for helping and motivating me in this work is very long and goes way beyond those mentioned on the dedication page. Likely the three people most important to this work are Tyler Reese, Brigitte Servatuis, and Randy Paffenroth. Tyler started working on Kirchhoff graphs shortly after he started at WPI as a PhD student in January 2014. Brigitte agreed to work with us because Tyler needed an advisor whose background included graph theory (mine at the time did not), and Randy joined in as more of an expert on linear algebra and related computations. Tyler quickly became the engine driving our work forward, sorting out and developing the good and bad ideas that the other three of us threw at him weekly.

Also crucial to this work were the efforts of three WPI undergraduates, who completed their Major Qualifying Projects (MQPs) at WPI under my supervision or cosupervision: Marcel Gietzmann-Sanders helped develop a linear-programming alogrithm and wrote a code to find individual but perhaps larger Kirchhoff graphs. Jessica Kejia Wang helped develop uniformity algorithm and wrote a code to find all of the Kirchhoff graphs with up to a certain number of edge vectors. Finally, Grace Baumgartner worked on understanding how several reaction networks can be described in terms of Kirchhoff graphs. All three made significant contributions to this work.

https://doi.org/10.1515/9783111408576-203

In terms of the production of this monograph, I wish to thank Evan Sayer (currently my PhD student) for proof reading an early complete draft manuscript, Steven Elliot, then Senior Mathematics Editor, who helped with my book proposal at De Gruyter, and Ute Skambraks, the STM Content Editor at De Gruyter, who helped with the details of organizing this manuscript. Finally I wish to thank Vilma Vaičeliūnienė who as production manager was responsible for preparing the manuscript for printing.

Berlin, Germany, June 2025   Joseph D. Fehribach

Notes to the reader

The readers should be aware of the following points as they work through this book:

1. The count of the edge vectors s_i and vertices v_i (when the vertices are numbered) normally begins with 1, except when the first edge vector or vertex is distinguished for some purpose. Then this initial element will start with 0. This will generally be the case where $s_0 = b$ is the overall reaction for some reaction network.

2. Through out this monograph, G denotes some standard graph, D denotes some digraph, and $\boldsymbol{G}$ is a vector graph, perhaps a Kirchhoff graph, perhaps not.

3. For vector graphs and particularly Kirchhoff graphs over the rationals, vertices are in blue, and for relatively small multiplicities, edge vector multiplicities greater than one are indicated by red hash marks. When multiplicities are larger, they are given as red integers.

4. **Notation**: Some letters are used differently in different places in this text, for example i and j are used as different indices from section to section. The following letters are the most important to be used consistently throughout the text:
   - A: any matrix row-equivalent to R
   - B: incidence matrix (row-equivalent to R)
   - R: row matrix
   - N: null matrix
   - S: finite set of edge vectors $\{s_0, s_1, \ldots, s_n\}$
   - n: total number of edge vectors
   - k: number of linearly independent edge vectors
   - m: edge vector multiplicity for a uniform vector or Kirchhoff graph, *except in Chapter* 5
   - m^*: minimum edge vector multiplicity (for a uniform Kirchhoff graph)

https://doi.org/10.1515/9783111408576-204

Contents

1 Introduction

This introduction to Kirchhoff graphs is given in two parts: first from a mathematical view, and then from an electrochemistry view. Though the mathematical view comes first, the two are independent, and the reader may choose which one to read first. Historically, Kirchhoff graphs were developed in electrochemistry, though they can be defined and discussed independently of this history and can be applied to any reaction network or networking process where the stoichiometry is known or at least something of interest. Perhaps the key impetus for considering Kirchhoff graphs comes from their connection to reaction networks: they are essentially circuit diagrams for these networks.

1.1 Mathematical side

Perhaps it is best to start a discussion on Kirchhoff graphs with a simple example.

Example 1.1. Consider these four vectors in $\mathbb{R}^2$:

$$s_0 = \begin{bmatrix} 2 \\ 0 \end{bmatrix}, \quad s_1 = \begin{bmatrix} 0 \\ 2 \end{bmatrix}, \quad s_2 = \begin{bmatrix} 1 \\ 1 \end{bmatrix}, \quad s_3 = \begin{bmatrix} 1 \\ -1 \end{bmatrix}$$

and let $S = \{s_0, s_1, s_2, s_3\}$ be the set of these vectors. Notice that $s_0 + s_1 - 2s_2 = 0$ and also that $s_0 - s_1 - 2s_3 = 0$. These dependencies are depicted in the vector graph[1] in Figure 1.1, where the edges are the vectors s_i. In addition, notice that the columns of the null matrix

$$N = \begin{bmatrix} 1 & 1 \\ 1 & -1 \\ -2 & 0 \\ 0 & -2 \end{bmatrix}$$

also express these dependencies; the entries in each column give the coefficients of the four vectors in the two linear combinations given above that sum to zero. Thus:

$$\begin{bmatrix} s_0, & s_1, & s_2, & s_3 \end{bmatrix} N = \begin{bmatrix} s_0, & s_1, & s_2, & s_3 \end{bmatrix} \begin{bmatrix} 1 & 1 \\ 1 & -1 \\ -2 & 0 \\ 0 & -2 \end{bmatrix} = \begin{bmatrix} 0, & 0 \end{bmatrix}$$

The standard orthogonal complement matrix of N is the matrix

$$R = \begin{bmatrix} 2 & 0 & 1 & 1 \\ 0 & 2 & 1 & -1 \end{bmatrix} \tag{1.1}$$

---

1 A vector graph is a graph whose edges are vectors, or whose edges are assigned vectors based on an associated directed graph. Vector graphs are formally defined below.

https://doi.org/10.1515/9783111408576-001

meaning that the product of R and N is zero:

$$RN = \begin{bmatrix} 0 & 0 \\ 0 & 0 \end{bmatrix}$$

and the transpose of the rows of R and columns of N form a basis for $\mathbb{R}^4$. This result is known either as the fundamental theorem of linear algebra or the row-space, null-space orthocomplementarity [29, pp. 185, 198]. Also notice that the columns of R are exactly the four vectors of S, so indeed:

$$R = [s_0, \quad s_1, \quad s_2, \quad s_3] = \begin{bmatrix} 2 & 0 & 1 & 1 \\ 0 & 2 & 1 & -1 \end{bmatrix}$$

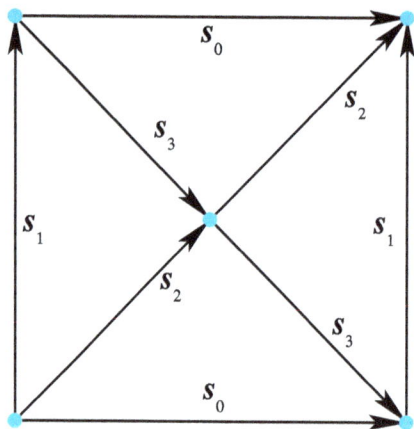

Figure 1.1: An example of a Kirchhoff graph for the vectors in the set S in Example 1.1. Vertices are in blue. Notice that $s_0 + s_1 - 2s_2$ and $s_0 - s_1 - 2s_3$ are cycles in this graph, while the cut vector (incidence vector) for each vertex is either [1, 1, 1, 0], [1, −1, 0, 1], the negative of one of these two, or the zero vector.

Notice that the vertex cuts[2] (incidence vectors) of this vector graph all correspond to vectors in Row(R)—the row space of R—the space of all linear combinations of the rows of R. In general, if an edge vector originates from (exits) a vertex, then its entry in the cut vector is +1; if an edge vector terminates at (enters) a vertex, then its entry in the cut vector is −1; if an edge vector is not incident on a vertex or if it enters and leaves a single vertex, then its entry in the cut vector is 0. In this specific example, for the vertex at the lower left, the vertex cut corresponds to half the sum of the two rows of R, while for the vertex at the upper left, the vertex cut corresponds to half the difference of the

---

2 The *vertex cut* or *incidence vector* for a given vertex is the vector whose i-th entry is the number of copies of the edge vector s_i that exits the vertex minus the number of copies that enter the vertex.

first row minus the second row. The vertex cuts for the other two corner vertices are the negatives of the first two, whereas the vertex at the center is a null vertex, whose vertex cut is the zero vector. Thus all five vertices of the vector graph in Figure 1.1 "lie" in Row(R), and indeed these vertex cuts span this row space.

The same sort of correspondence exists between the cycles (or closed walks, or circuits) of our vector graph and the Col(N)—the column space of N—the space of all linear combinations of the columns of N. Recall of course that Col(N) = Null(R)—the null space of R. Along with the two linear combinations of the edge vectors mentioned above, one can check that all other closed walks correspond to vectors in Col(N) and Null(R) as well. Indeed, there is a cycle basis in our Kirchhoff graph in Figure 1.1 corresponding to any integral basis in Null(R), including the columns of N.

That the cycles (closed walks, circuits) of the vector graph in Figure 1.1 all correspond to vectors in Null(R) = Col(N) while all its vertex cuts "lie" in Row(R) is the key defining condition making this vector graph a Kirchhoff graph for the four vectors in $S =$ $\{s_0, s_1, s_2, s_3\}$. The same is true for any set of four vectors having exactly the same dependencies as these four. So this vector graph in Figure 1.1 is a Kirchhoff graph for the row matrix R or for any matrix A that is row-equivalent to R since the columns of R and the columns of A have exactly the same dependencies as the vectors in S. Notice that A will necessarily have the same number of columns n as does R, though the number of rows of A could be any number greater than or equal to k. In addition, notice that the *incidence matrix*[3] B for our Kirchhoff graph is one such matrix A—the incidence matrix is necessarily row-equivalent to R. For this Kirchhoff graph in Figure 1.1, the incidence matrix is

$$B = \begin{bmatrix} 1 & 1 & 1 & 0 \\ -1 & 1 & 0 & -1 \\ 0 & 0 & 0 & 0 \\ 1 & -1 & 0 & 1 \\ -1 & -1 & -1 & 0 \end{bmatrix}$$

where the choice of how the vertices are indexed (here, bottom to top, and inside that, from left to right) is made so that the incidence matrix B is vertically antisymmetric about the middle row. Notice that unlike for a standard digraph, although the column entries still always sum to zero, there may be more than two nonzero entries in each column. Having more than two nonzero entries is due to the same edge vector appearing in multiple places in the vector graph.

One might wonder if Kirchhoff graphs are unique: Is the Kirchhoff graph in Figure 1.1 the only Kirchhoff graph for the vectors of S? The answer is in fact "No." The

---

3 In the incidence matrix for a vector graph, the rows correspond to vertices, and the columns correspond to edge vectors. In the incidence matrix B, the absolute value of the entry b_{ij} is the net number of copies of edge vector s_j incident on vertex v_i; the entry is positive if these edge vectors exit the vertex and negative if they enter the vertex.

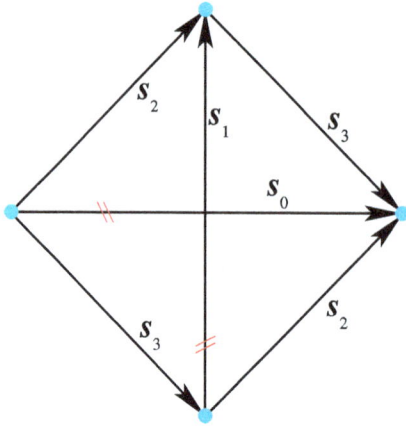

Figure 1.2: A second example of a Kirchhoff graph for the vectors in the set S in Example 1.1. Red hash marks indicate two copies of s_1 between the left and right vertices and two copies of s_2 between the bottom and top vertices. The edge vector crossing in the center of the graph is *not* a vertex, but because of the vector nature of the edges, this crossing is unavoidable.

vector graph in Figure 1.2 is also a Kirchhoff graph for S or R or any matrix row-equivalent to R, including the incidence matrix for this Kirchhoff graph. Its vertex cuts all correspond to vectors that span Row(R), and its cycles all correspond to vectors that span Null(R). To accomplish this result, there must be two copies of s_0 connecting the left-most and right-most vertices, and two copies of s_1 connecting the bottom and top vertices. The number of copies of a vector connecting two vertices is the multiplicity (weight) for each of these edge vectors. The Kirchhoff graph in Figure 1.1 is called the *Square* and the Kirchhoff graph in Figure 1.2 is called the *Diamond*. Later it will be shown that these two are in fact the only two Kirchhoff graphs for this set of edge vectors S having two or fewer copies of each edge vector.

It should be mentioned that although the standard (canonical) representation of the edge vectors s_i is often convenient, it is the dependencies that define a Kirchhoff graph, not the particular representation chosen for edge vectors.

Example 1.2. Consider the following row-matrix/null-matrix pair:

$$R = \begin{bmatrix} 1 & 0 & 1 \\ 0 & 1 & -1 \end{bmatrix}, \quad N = \begin{bmatrix} 1 \\ -1 \\ -1 \end{bmatrix}$$

Several projections of a Kirchhoff graph for this pair are shown in Figure 1.3. The first uses the canonical vectors:

$$s_1 = \begin{bmatrix} 1 \\ 0 \end{bmatrix}, \quad s_2 = \begin{bmatrix} 0 \\ 1 \end{bmatrix}, \quad s_3 = \begin{bmatrix} 1 \\ -1 \end{bmatrix}$$

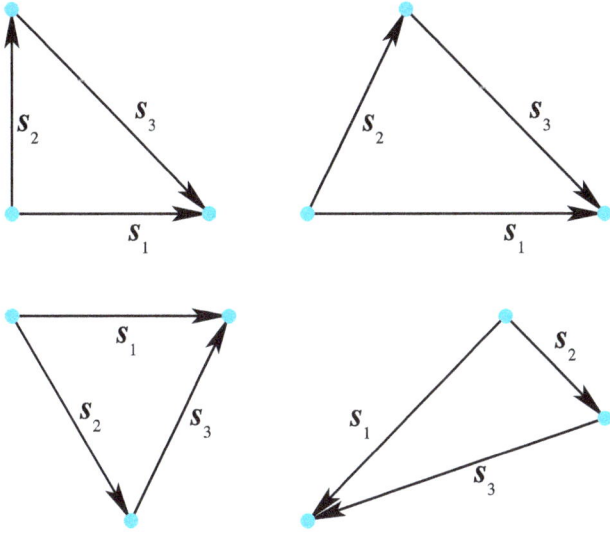

Figure 1.3: Four representations of the same Kirchhoff graph. The upper left graph is the canonical representation using the columns of R. The other three show the same cycle with the same vertex cuts.

but the other three are equally representations of the same Kirchhoff graph since they represent the same cycle and the same vertex cuts. There are, of course, many other representations.

1.1.1 Deeper on the mathematical side

Perhaps the most important consequence of these first two examples are that they are in no way one-offs, but rather generalize to all Kirchhoff graphs. What follows is a discussion of the basic mathematical ideas for Kirchhoff graphs in general, not tied to the specific example above.

Let $\mathcal{V}$ be any vector space, and consider the subspace spanned by the set $\mathcal{S} := \{s_1, s_2, \ldots, s_n\} \subset \mathcal{V}$ for some finite $n \in \mathbb{Z}^+$ (positive integers). Suppose that the vectors in $\mathcal{S}$ form only *rational* linear combinations, and for simplicity, suppose that no vector in $\mathcal{S}$ is a scalar multiple of another. Finally, suppose that there is an integer k with $1 < k < n$ such that $\{s_1, s_2, \ldots, s_k\}$ forms a basis for span($\mathcal{S}$). It is then possible to write the vectors $\{s_{k+1}, s_{k+2}, \ldots, s_n\}$ as linear combinations of this basis.

Let $c'_{i,j} \in \mathbb{Q}$ be the unique coefficients such that:

$$
\begin{aligned}
s_{k+1} &= c'_{1,1}s_1 &+& \quad c'_{2,1}s_2 &+& \quad \cdots &+& \quad c'_{k,1}s_k \\
s_{k+2} &= c'_{1,2}s_1 &+& \quad c'_{2,2}s_2 &+& \quad \cdots &+& \quad c'_{k,2}s_k \\
&\vdots \quad \vdots & & \qquad \vdots \\
s_n &= c'_{1,n-k}s_1 &+& \quad c'_{2,n-k}s_2 &+& \quad \cdots &+& \quad c'_{k,n-k}s_k
\end{aligned}
$$

Let $q \in \mathbb{Z}^+$ be the least common multiple of denominators of these rational coefficients: $q := \mathrm{lcm}(\mathrm{dnm}(c'_{i,j}))$, $\forall\, i, j$, $1 \leq i \leq k$, $1 \leq j \leq n - k$, and define $c_{i,j} := q c'_{i,j}$. These new coefficients are all defined to be integers, $c_{i,j} \in \mathbb{Z}$, and our system can be written as:

$$
\begin{aligned}
q s_{k+1} &= c_{1,1} s_1 + c_{2,1} s_2 + \cdots + c_{k,1} s_k \\
q s_{k+2} &= c_{1,2} s_1 + c_{2,2} s_2 + \cdots + c_{k,2} s_k \\
&\ \vdots \qquad\qquad\quad \vdots \\
q s_n &= c_{1,n-k} s_1 + c_{2,n-k} s_2 + \cdots + c_{k,n-k} s_k
\end{aligned}
\tag{1.2}
$$

This system of equations can be rearranged as:

$$
\begin{aligned}
c_{1,1} s_1 + c_{2,1} s_2 + \cdots + c_{k,1} s_k - q s_{k+1} &= 0 \\
c_{1,2} s_1 + c_{2,2} s_2 + \cdots + c_{k,2} s_k \qquad\ - q s_{k+2} &= 0 \\
\vdots \qquad\qquad\qquad\qquad\qquad \vdots \qquad\ \vdots \\
c_{1,n-k} s_1 + c_{2,n-k} s_2 + \cdots + c_{k,n-k} s_k \qquad\qquad\qquad - q s_n &= 0
\end{aligned}
$$

Now consider the coefficient matrix defined using the integer coefficients:

$$
C := \begin{bmatrix}
c_{1,1} & c_{1,2} & \cdots & c_{1,n-k} \\
c_{2,1} & c_{2,2} & \cdots & c_{2,n-k} \\
\vdots & \vdots & & \vdots \\
c_{k,1} & c_{k,2} & \cdots & c_{k,n-k}
\end{bmatrix}
$$

Consider also a larger matrix N formed by placing the matrix C in an upper block over the negative of the appropriately sized identity block multiplied by q:

$$
N := \begin{bmatrix} C \\ -q I_{n-k} \end{bmatrix} = \left[
\begin{array}{cccc}
c_{1,1} & c_{1,2} & \cdots & c_{1,n-k} \\
c_{2,1} & c_{2,2} & \cdots & c_{2,n-k} \\
\vdots & \vdots & & \vdots \\
c_{k,1} & c_{k,2} & \cdots & c_{k,n-k} \\
\hline
-q & 0 & \cdots & 0 \\
0 & -q & \cdots & 0 \\
\vdots & \vdots & & \vdots \\
0 & 0 & \cdots & -q
\end{array}
\right]
$$

Based on the system of vector equations (1.2), notice that $[s_1, s_2, \ldots, s_n]$, the row vector of the vectors of S, is in a sense in the left null space of the matrix N:

$$
[s_1, s_2, \ldots, s_n] N = [s_1, s_2, \ldots, s_n] \begin{bmatrix} C \\ -q I_{n-k} \end{bmatrix} = [0, 0, \ldots, 0]
\tag{1.3}
$$

Equation (1.3) shows that N is the null matrix for our set of vectors S. The columns of N give a basis for the dependencies for S (these columns are clearly linearly independent, and they must span the dependencies, or our set of vectors S cannot have a basis). Now define the row matrix R that is the orthogonal complement of N:

$$R := [qI_k \mid C] = \begin{bmatrix} q & 0 & \cdots & 0 & c_{1,1} & c_{1,2} & \cdots & c_{1,n-k} \\ 0 & q & \cdots & 0 & c_{2,1} & c_{2,2} & \cdots & c_{2,n-k} \\ \vdots & \vdots & & \vdots & \vdots & \vdots & & \vdots \\ 0 & 0 & \cdots & q & c_{k,1} & c_{k,2} & \cdots & c_{k,n-k} \end{bmatrix}$$

Thus $RN = [0]$, where $[0]$ is the $k \times (n-k)$ zero matrix, and R is maximal in the sense that there are no row vectors outside the row space $\text{Row}(R)$ that lie in the left null space of N.

Now suppose that A is *any* matrix that is row-equivalent to R. Then $\text{Row}(A) = \text{Row}(R)$, and by the fundamental theorem of linear algebra the $\text{Null}(A) = \text{Null}(R)$ is the orthogonal complement of this row space. Indeed, the rows of R are a basis for $\text{Row}(A)$, and the columns of N are a basis for $\text{Null}(A)$. In addition, since $[s_1, s_2, \ldots, s_n]N = [0, 0, \ldots, 0]$, $RN = [0]$, and $AN = [0]$, the columns of R and A have the same dependencies as the vectors in S. Hence the columns of either R or A can be used to represent the vectors in S regardless of which vector space V one started with; this result is basically an implication of the fact that all finite-dimensional vector spaces of a given dimension are isomorphic. The columns of R can be taken as the *canonical representation* of the vectors in S regardless of which vector space our edge vectors come from, though at times, the columns of some other row-equivalent matrix A will be a more convenient representation for the vectors of S.

What does this have to do with Kirchhoff graphs? Suppose that a graph can be constructed whose edges are vectors, say the columns of R, containing all of the cycles, circuits, and closed walks corresponding to the columns of N, no other cycles or closed walks, and each vertex cut lies in $\text{Row}(R)$. Such a vector graph is a *Kirchhoff graph* for S and for R. Before giving a formal definition for *Kirchhoff graph*, a definition for *vector graph* is needed. Informally, a vector graph is a graph whose edges are vectors. A formal definition is next.

Definition. Given a vector space V, a finite set of vectors $S = \{s_1, s_2, \ldots, s_n\} \subset V$, and a finite set of vertices V, a **vector graph** $G := G(V, S)$ is a set of ordered quadruples $(v_i, s_\ell, v_j, m_{i,j}) \in V \times S \times V \times \mathbb{N}$ subject to the **consistency condition** that there is a cycle, circuit, or closed walk in the graph if and only if the corresponding linear combination of the edge vectors in a walk add to zero in the vector space V. The edge vector s_ℓ exits vertex v_i and enters vertex v_j. The nonnegative integer $m_{i,j}$ is the multiplicity of the edge vector incident on v_i and v_j. Vectors may be traversed in either direction in a cycle, circuit, or walk. Any number of copies of a single vector may exit one vertex and enter another vertex, but two copies of a single vector cannot exit a single vertex and enter distinct vertices, or exit distinct vertices and enter a single vertex. If $v_i = v_j$, then $s_\ell = \mathbf{0}$.

Remarks.

1. The natural space to embed a vector graph G is $\mathbb{R}^k$, where k is the number of linearly independent vectors in S, but G can be drawn in a lower-dimensional space (often the Euclidean plane) as long as the consistency condition is met.

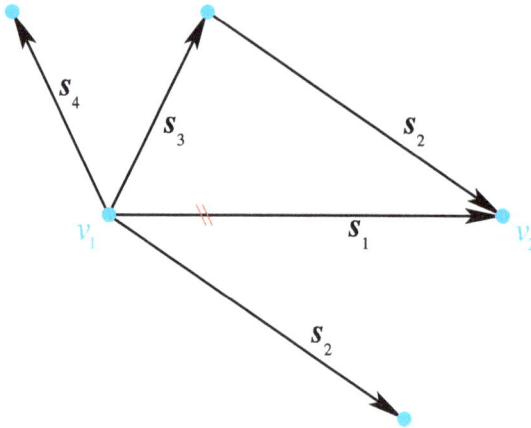

Figure 1.4: A simple example of a vector graph. The cycle in this graph corresponds to $s_1 - s_2 - s_3 = \mathbf{0}$. Note that two copies of s_1 exit v_1 and enter v_2 as indicated by the hash marks.

2. An example of a vector graph is shown in Figure 1.4. Note that the cycle in the graph implies that the corresponding vectors must add to zero in the appropriate vector space, taking the orientation of the vectors into account.

3. Suppose that $S = \{\sin^2 x, \cos^2 x, 1\} \subset L^2[0, 2\pi]$. Suppose that there are three vertices v_1, v_2, and v_3, that $\sin^2 x$ exits v_1 and enters v_2, and that $\cos^2 x$ exits v_2 and enters v_3. If 1 now exits v_1, then the only possibility for this to be a vector graph is that it enters v_3. Indeed the resulting vector graph is a Kirchhoff graph.

Definition. Given a vector space V and a finite set of vectors $S = \{s_1, s_2, \dots, s_n\} \subset V$, let the matrices R and N be defined as above, and suppose that G is a *cyclic* vector graph using vectors from S. If the columns of N correspond to a basis for the cycle space of G, and if the vertex cuts of G correspond to vectors in Row(R), then G is a **Kirchhoff graph** for S and R and any matrix A that is row-equivalent to R.

Remarks.

1. The definition of the term *cyclic* is a bit involved, so it is deferred until Section 2.3. For the moment, think of it as implying that the vector graph has no vertices of de-

gree one, that is, no twigs. The complete definition does imply this, but also implies more.

2. One of the interesting possibilities to consider is what happens if the row space Row(R) and the null space Null(R) are interchanged: Suppose the vertex cuts in a vector graph span Null(R) while the cycles in this graph span Row(R). Such a vector graph would also be a Kirchhoff graph, but this Kirchhoff graph would be the *Kirchhoff dual* of **G** in our definition. Kirchhoff duals along with the related concept of Maxwell reciprocal diagrams will be discussed in Chapter 7 (Kirchhoff graph duality and Maxwell reciprocal diagrams).

3. It is worth noting here that for vector graphs, the cycle space and the cut space are modules, not vector spaces. The issue is that there need to be integral numbers of each vector in a cycle or cut, and unlike for standard graphs and digraphs, finite fields are not the complete answer. Using finite fields is discussed in Chapter 5 (Matroids and Kirchhoff graphs over finite fields).

1.2 Electrochemistry side

As was mentioned at the beginning of this Introduction, Kirchhoff graphs are essentially circuit diagrams for reaction networks, including chemical and electrochemical reaction networks. In this context, Kirchhoff graphs are graphical representations of reaction network stoichiometry. Cycles in Kirchhoff graphs represent the conservation of all the species in the reaction network and therefore also the electrochemical potential[4] (this corresponds to the Kirchhoff voltage law), whereas vertex cuts in Kirchhoff graphs represent the conservation of reaction rates (what flows in, flows out; this corresponds to the Kirchhoff current law). Notice that these rates are in effect currents—rates of flow. Mathematically this means that there are cycles forming a basis for the null space of the stoichiometric matrix for the reaction network, whereas the vertex cuts all correspond to vectors in the row space of this stoichiometric matrix. Thus Kirchhoff graphs satisfy a version of the Kirchhoff laws in much the same ways as do electrical circuit diagrams, and the example presented next will indicate how.

Example 1.3 (Hydrogen evolution reaction (HER)). One relatively simple reaction network that can be used to illustrate the concept and usefulness of a Kirchhoff graph is the HER (hydrogen evolution reaction) network. This network is important in, among other places, certain corrosion processes. There is a single overall reaction with three reaction steps:

---

4 This can be generalized to a nonconservative system where energy is required to move around a cycle. In this case, species are still conserved around a cycle, but the electrochemical potential is not.

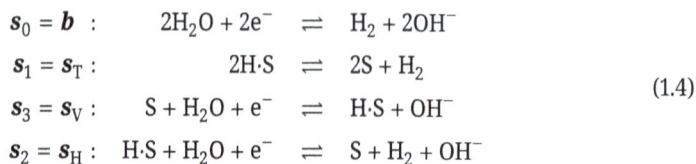

$$
\begin{aligned}
s_0 = b \;:& \qquad 2H_2O + 2e^- \;&\rightleftharpoons&\; H_2 + 2OH^- \\
s_1 = s_T \;:& \qquad 2H\cdot S \;&\rightleftharpoons&\; 2S + H_2 \\
s_3 = s_V \;:& \quad S + H_2O + e^- \;&\rightleftharpoons&\; H\cdot S + OH^- \\
s_2 = s_H \;:& \quad H\cdot S + H_2O + e^- \;&\rightleftharpoons&\; S + H_2 + OH^-
\end{aligned}
\tag{1.4}
$$

where H_2 (molecular hydrogen), OH^- (peroxide ion), H_2O (water), and e^- (electron) are the terminal species that appear in the overall reaction, while S (a reaction site on the catalyst surface) and H·S (a hydrogen atom attached to such a site) are the intermediate species that do not appear in the overall reaction. The concentrations of the intermediate species are taken to be constant as the overall reaction moves forward (or backward). The numbering of the steps $\{0, 1, 2, 3\}$ matches that of the Kirchhoff graphs in Example 1.1. The letter b replaces s_0 and denotes the overall reaction because of a convention in mathematics since it is achieved as a linear combination of the three reaction steps ($Ax = b$). The subscripts that distinguish the steps honor, respectively, Tafel, Volmer, and Heyrovsky, three famous electrochemists. These three steps are shown schematically in Figures 1.5–1.7. Keep in mind that the steps are reversible, so the choice of the forward

Figure 1.5: Tafel reaction step. On the left, two hydrogen atoms are attached to a catalytic surface; on the right, a hydrogen molecule is near two open sites on the surface. The reaction moves hydrogen back and forth between these two states.

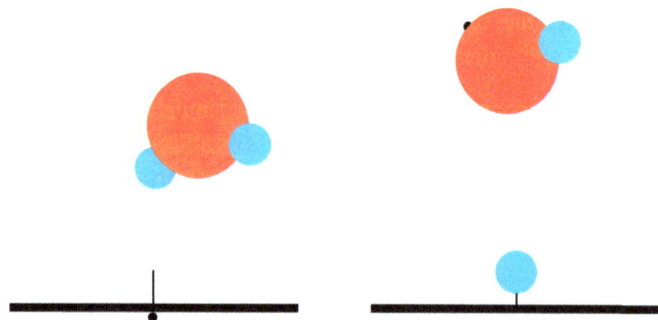

Figure 1.6: Volmer reaction step. On the left, a water molecule is near an open site on a catalytic surface; on the right, a hydrogen atom is attached to the surface with a hydroxide ion near this surface. That an electron moves between the surface and the hydroxide ion in the bulk solution implies an electrical current. Again the reaction moves back and forth between these two states.

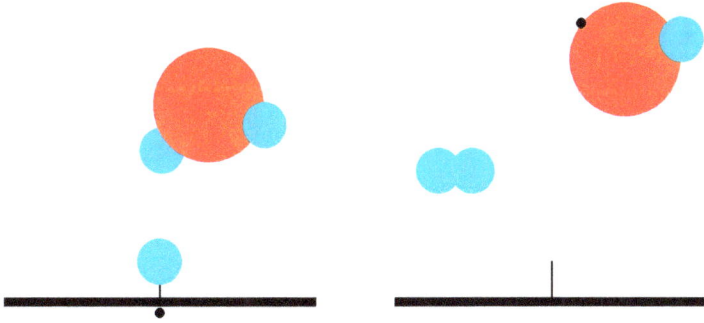

Figure 1.7: Heyrovsky reaction step. On the left, a water molecule is near a hydrogen atom attached to a catalytic surface; on the right, a hydrogen molecule and a hydroxide ion are near an open site on the surface. Again, an electron moving between the surface and the hydroxide ion in the bulk solution implies an electrical current. The reaction moves back and forth between these two states.

Table 1.1: Stoichiometric coefficients for the hydrogen evolution reaction (HER).

	H_2	OH^-	H_2O	e^-	S	H·S
b	1	2	−2	−2	0	0
s_T	1	0	0	0	2	−2
s_V	0	1	−1	−1	−1	1
s_H	1	1	−1	−1	1	−1

direction is somewhat arbitrary. In the Tafel step, for example, two hydrogen atoms are together on the surface on one side, whereas a hydrogen molecule is near two open sites on the other side. The Tafel step moves back and forth between these two arrangements. The other two steps behave similarly.

Moving the reactant coefficients to the right-hand side and giving them negative signs, one obtains the stoichiometric coefficients table in Table 1.1. Notice that overall reaction and the reaction steps can be expressed as stoichiometric vectors using these stoichiometric coefficients:

$$
b = \begin{bmatrix} 1 \\ 2 \\ -2 \\ -2 \\ 0 \\ 0 \end{bmatrix}, \quad
s_T = \begin{bmatrix} 1 \\ 0 \\ 0 \\ 0 \\ 2 \\ -2 \end{bmatrix}, \quad
s_V = \begin{bmatrix} 0 \\ 1 \\ -1 \\ -1 \\ -1 \\ 1 \end{bmatrix}, \quad
s_H = \begin{bmatrix} 1 \\ 1 \\ -1 \\ -1 \\ 1 \\ -1 \end{bmatrix}
$$

Part of the question then is how can b be written as a linear combination of s_T, s_V, and s_H; these are the *reaction pathways* for this network. Based on these vectors, one can obtain the stoichiometric matrix (or its transpose in some parts of the literature):

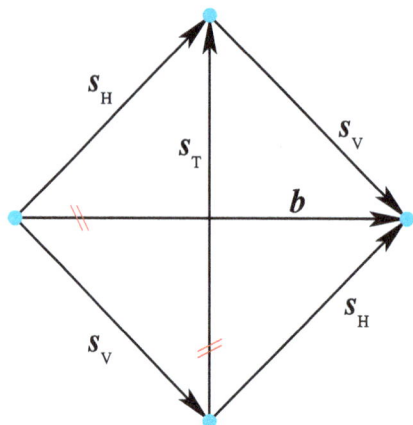

Figure 1.8: A Kirchhoff graph for the HER reaction network showing how the three steps (vectors) can combine (add) to yield the overall reaction **b**. Again the red hash marks indicate that two copies of s_T and **b** lie in parallel between two pairs of vertices. Notice that this Kirchhoff graph is identical to the one in Figure 1.2.

$$A = \begin{bmatrix} 1 & 1 & 0 & 1 \\ 2 & 0 & 1 & 1 \\ -2 & 0 & -1 & -1 \\ -2 & 0 & -1 & -1 \\ 0 & 2 & -1 & 1 \\ 0 & -2 & 1 & -1 \end{bmatrix} \tag{1.5}$$

For this network (these reaction steps and this stoichiometric matrix), a Kirchhoff graph is shown in Figure 1.8. There are two main features of this Kirchhoff graph that both define it as a Kirchhoff graph for this reaction network and make it useful in understanding this reaction network. The first is that all the reaction pathways combining the three steps s_T, s_V, and s_H to yield the overall reaction **b** can be seen in the graph. Without regard to order (meaning that $s_V + s_H$ is seen in the same way as $s_H + s_V$) and not counting combinations that form cycles among the three steps, there are three combinations of s_T, s_V, and s_H that yield **b**: (1) $s_V + s_H$, (2) $s_V + s_T + s_V$, and (3) $s_H - s_T + s_H$. The minus sign indicates that the Tafel step occurs in the reverse direction in the third pathway. Thus for (1), $s_V + s_H = b$, or $-b + s_V + s_H = 0$, or equivalently, the cycle vector $[-1, 0, 1, 1]^T$ is in the null space of A:

$$\begin{bmatrix} 1 & 1 & 0 & 1 \\ 2 & 0 & 1 & 1 \\ -2 & 0 & -1 & -1 \\ -2 & 0 & -1 & -1 \\ 0 & 2 & -1 & 1 \\ 0 & -2 & 1 & -1 \end{bmatrix} \begin{bmatrix} -1 \\ 0 \\ 1 \\ 1 \end{bmatrix} = \begin{bmatrix} 0 \\ 0 \\ 0 \\ 0 \\ 0 \\ 0 \end{bmatrix} = 0$$

The same is true for the other two combinations: for (2), the cycle vector is $[-1, 1, 2, 0]^T$, and for (3), the cycle vector is $[-1, -1, 0, 2]^T$, and both of these are also in the null space of A. Indeed, any two of these three form a basis[5] for this null space, and this is the first defining property for a Kirchhoff graph: the cycle vectors for the reaction pathways can be used to form a basis for the null space of A. This is also a generalization of the Kirchhoff voltage law: all of the reaction species are conserved around any cycle, and if all of the reaction steps are reversible, then the electrochemical potential for the system is conserved.

This Kirchhoff graph also shows that the HER network is robust in the sense that only two of the three steps are needed to carry out the overall reaction. Stopping exactly one of the steps by controlling, say, pH or temperature will not prevent the overall reaction from occurring because each pathway avoids one of the reaction steps: s_T is not in (1), s_H is not in (2), and s_V is not in (3). Stopping two of the steps, however, would prevent the overall reaction from occurring, at least via this reaction network.

The second defining feature is that at each vertex, the reaction rates for each of the steps incident on that vertex must sum to zero. So at the top vertex, $r_V - r_H - 2r_T = 0$. Here minus signs indicate that the associate steps head into this vertex, and the -2 means that two copies of the Tafel step head into this vertex. The bottom vertex is just the negative of the top vertex. From the reaction steps (1.4), one sees that the rate at which occupied sites H·S are produced or consumed is the negative of the rate at which open sites S are produced or consumed, and these are each given by the Volmer rate minus the Heyrovsky rate minus twice the Tafel rate. In symbols, $r_{H·S} = -r_S = r_V - r_H - 2r_T$, and from the rate balance indicated by the top or bottom vertex, $r_{H·S} = -r_S = 0$, meaning that the concentrations of the intermediate species H·S and S are constant as the overall HER process proceeds. In general, the rate balances at vertices not involving the overall reaction represent that the concentrations of intermediate species are constant as the overall reaction occurs in either the forward of backward reaction.

Again from the top (or bottom) vertex in the graph, $2r_T = r_V - r_H$, meaning that it is a sort of Wheatstone bridge: if the rates of the Volmer and Heyrovsky steps are equal, then the Tafel rate is zero, and the overall reaction rate is equal to that of the Volmer and Heyrovsky steps. On the other hand, if the Tafel rate is nonzero, then the Volmer and Heyrovsky rates differ by exactly twice this Tafel rate; measuring the rate and direction of the Tafel flow yields the difference between the Volmer and Heyrovsky rates, much as in the standard Wheatstone bridge experiment from an elementary physics course.

The same rate balances occur at the other vertices, and in general, each vertex cut lies in the row space of the stoichiometric matrix A. So for the top vertex, the vector $[0, -2, 1, -1]$ is the bottom row of A, for the left most vertex, the vector $[2, 0, 1, 1]$ is the second row of A, and so forth. This is a generalization of the Kirchhoff current law. There

---

5 That is, any of these three cycle vectors can be written as a linear combination of the other two. For example, if one adds the second and third cycle vectors, one obtains twice the first cycle vector.

is one important caveat, however, about the reaction rates: If the rates for steps s_H and s_V are both positive, then the overall reaction occurs in the forward direction. Nonetheless, the rate balance $r_V + r_H + 2r_b = 0$ implies that r_b is *negative*. The actual overall reaction rate is the negative of r_b:

$$r_{H_2} = r_{OH^-}/2 = r_{OR} = -r_b = (r_V + r_H)/2$$

All of the above information about the network is equally present in the list of the reaction steps, or in the stoichiometric matrix, as well as in the Kirchhoff graph; the key point, however, is that the information is seen more easily in the Kirchhoff graph. Mathematically, that this Kirchhoff graph gives a complete depiction of all the reaction pathways with the rate balances at each of the vertices is a consequence of the ortho-complementarity of the row space and the null space of the stoichiometric matrix A. This result is sometimes called the fundamental theorem of linear algebra [29, pp. 185, 198].

Finally, it is worth considering briefly a second closely related reaction network, the hydrogen oxidation reaction (HOR) network. As we will discuss in Section 8.2, the HOR network has the same stoichiometric matrix as A given in (1.5). This means that the null space and the row space for the HER and HOR networks are the same and hence that the Kirchhoff graph in Figure 1.8 will work for both networks. Having a common Kirchhoff graph indicates that in terms of stoichiometry, the HER and HOR reaction networks have the same structure and fundamentally work the same way. All of this is discussed in more detail in Chapter 8 (Kirchhoff graphs and reaction networks).

In general, Kirchhoff graphs for any reaction network will help someone under-stand their pathways and stoichiometric constraints. All the reaction pathways allowed by the stoichiometry will be present in the Kirchhoff graph, and the Kirchhoff graph vertices represent all the rate balances dictated by the stoichiometry. It is possible to determine whether controlling certain reaction steps will control the overall reaction, or whether knowing the reaction rates of certain reaction steps will determine certain or all other reaction rates. All of this is guaranteed by the mathematics of linear algebra. What is required is that the stoichiometry of the reaction steps is known, or at least one has some idea of what the steps are and wants to understand the implications of the sto-ichiometry. If the reaction steps are not known, then one may try out sets of proposed steps to see if their stoichiometry is consistent with what is known about the reaction network. Again these implications will be explored for other networks in Chapter 8.

1.3 Electrical circuits

At the beginning of this introduction, it was mentioned that the name *Kirchhoff graph* comes from the fact that these graphs are a generalization of the Kirchhoff laws applied to electrical circuits. Although a theorem establishing this equivalence can be stated and proven, perhaps the following simple example is more helpful.

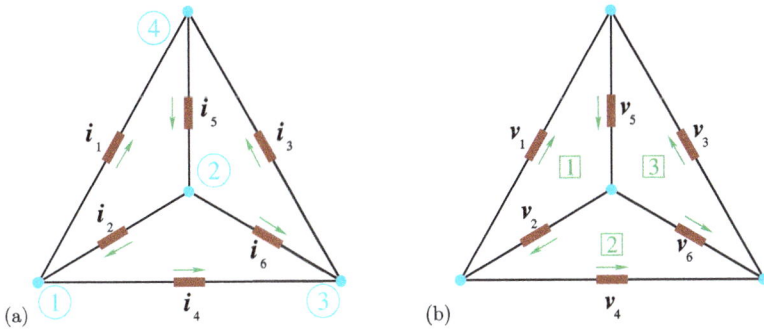

Figure 1.9: A circuit diagram for a simple electrical network. The nodes or vertices are in blue; the black line segments or edges are the electrical connections; the brown rectangles indicate elements (resistors, capacitors, inductors, devices, etc.), where the voltage changes as a current flows through. (a) Diagram showing the currents in each segment; current i_j flows through segment j. The green arrows give the positive direction for current flow in each segment; the actual current can flow in either direction, positive or negative. Blue circled numbers index the nodes. (b) Diagram giving the voltage drops for each segment. Green squared numbers index the inner circuits or cycles in the diagram.

Example 1.4. Consider the electrical circuit diagram shown in Figure 1.9. Recall that the Kirchhoff current law states that the sum of all the currents flowing into or out of any node (vertex) must be zero. Applying this law to this particular example, one obtains a linear system:

$$
\begin{aligned}
i_1 - i_2 \quad\quad + i_4 \quad\quad\quad\quad &= 0 \\
i_2 \quad\quad\quad - i_5 + i_6 &= 0 \\
i_3 - i_4 \quad - i_6 &= 0 \\
-i_1 \quad - i_3 \quad + i_5 \quad &= 0
\end{aligned}
\tag{1.6}
$$

where each of the four equations comes from one of the four vertices. Of the four equations, only three are linearly independent. The coefficient matrix A for system (1.6) is

$$
A = \begin{bmatrix}
1 & -1 & 0 & 1 & 0 & 0 \\
0 & 1 & 0 & 0 & -1 & 1 \\
0 & 0 & 1 & -1 & 0 & -1 \\
-1 & 0 & -1 & 0 & 1 & 0
\end{bmatrix}
$$

and row operations lead to the reduced-echelon form matrix:

$$
R = [I|C] = \begin{bmatrix}
1 & 0 & 0 & 1 & -1 & 1 \\
0 & 1 & 0 & 0 & -1 & 1 \\
0 & 0 & 1 & -1 & 0 & -1
\end{bmatrix}
$$

and its orthogonal complement matrix:

$$N = [C/-I] = \begin{bmatrix} 1 & -1 & 1 \\ 0 & -1 & 1 \\ -1 & 0 & -1 \\ -1 & 0 & 0 \\ 0 & -1 & 0 \\ 0 & 0 & -1 \end{bmatrix}$$

Thus $Ai = 0$ and $Ri = 0$, where $i = [i_1\ i_2\ i_3\ i_4\ i_5\ i_6]^T$, and hence both are equivalent to system (1.6).

The columns of this matrix N should correspond to the voltage sums around the cycles for this circuit. This is indeed the case: Let v_i be the voltage change (positive or negative) associated with the ith edge segment on the right in Figure 1.9. The Kirchhoff voltage (potential) law states that the sum of all the voltage changes around any cycle must be zero. Applying this law to our example, we find a second linear system:

$$\begin{aligned}
v_1 + v_2 \qquad\qquad\quad + v_5 \qquad\qquad &= 0 \\
v_2 \qquad + v_4 \qquad\quad - v_6 &= 0 \\
v_3 \qquad\qquad + v_5 + v_6 &= 0 \\
-v_1 \qquad + v_3 + v_4 \qquad\qquad\qquad &= 0
\end{aligned} \tag{1.7}$$

As was the case for system (1.6), this voltage system can be written as a matrix equation and then reduced to a row-equivalent form. For system (1.7), the matrix equation $Dv = 0$ can be obtained, where $v = [v_1\ v_2\ v_3\ v_4\ v_5\ v_6]^T$ and

$$D = \begin{bmatrix} 1 & 0 & -1 & -1 & 0 & 0 \\ -1 & -1 & 0 & 0 & -1 & 0 \\ 1 & 1 & -1 & 0 & 0 & -1 \end{bmatrix}$$

This confirms that the columns of N also give the conditions of the Kirchhoff voltage law since $N = D^T$. This orthogonality condition is of course the key feature in the definition of Kirchhoff graphs.

In terms of electrical network terminology, the circuit matrix is essentially N^T, while the cutset matrix is essentially R. The columns of both N^T and R are indexed by the edges of the network graph, and since $RN = [0]$, its transpose $N^T R^T = [0]$. A network with positive resistors is known to be uniquely solvable if and only if the subset of edges of the network graph determined by the voltage sources does not contain a cycle, and the subset of edges determined by the current sources does not contain a cutset. Note, however, that the network graph is typically not recoverable from the matrices N and R. For a discussion of the connections between matroids and circuit diagrams, see, for example, Recski [22] (1989).

1.4 Examples of Kirchhoff graphs

This is the pretty-picture section of this book.[6] The next several chapters present a number of general results about Kirchhoff graphs and discuss how to construct such graphs. Before turning to this material, however, it would seem appropriate to present a few examples that make clear that Example 1.1 above is in no way unique, but rather one of a large set of Kirchhoff graphs.

Example 1.5. Consider the matrix

$$A = \begin{bmatrix} 3 & -2 & 1 & 4 & 2 \\ -2 & 1 & 3 & -3 & -5 \\ 1 & -2 & 7 & 0 & -6 \end{bmatrix} \tag{1.8}$$

which can be row reduced and presented in canonical form as the R, N-matrix pair:

$$R = \begin{bmatrix} 1 & 0 & 0 & 2 & 1 \\ 0 & 1 & 0 & 1 & 0 \\ 0 & 0 & 1 & 0 & -1 \end{bmatrix}, \quad N = \begin{bmatrix} 2 & 1 \\ 1 & 0 \\ 0 & -1 \\ -1 & 0 \\ 0 & -1 \end{bmatrix} \tag{1.9}$$

One possible Kirchhoff graph for both A in (1.8) and R in (1.9) is shown in Figure 1.10. Since the identity block of R in (1.9) is 3×3, this Kirchhoff graph naturally embeds in $\mathbb{R}^3$,

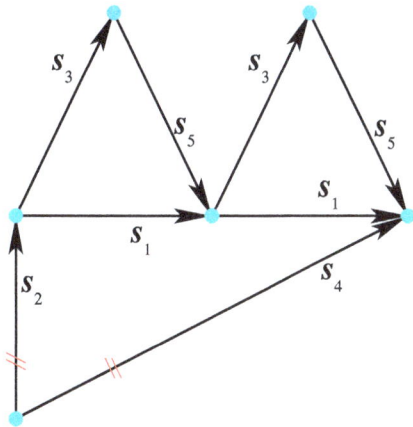

Figure 1.10: A Kirchhoff graph for A in (1.8) and R in (1.9). Vertices are again in blue; red hash marks indicate the number of copies of an edge vector connecting indicated vertices. Cycles, circuits or closed walks for this Kirchhoff graph correspond to vectors spanning Null(R) = Null(A), while all of the vertex cuts correspond to the vectors in Row(R) = Row(A). Each edge vector appears twice in this Kirchhoff graph.

6 Readers can decide for themselves whether or not the pictures (Kirchhoff graphs) are indeed pretty.

although it is relatively easy to draw in the plane. Still one can imagine the vectors s_1, s_2, and s_3 coming together at right angles at the middle-left vertex. The triangular cycles correspond directly to the two columns of N in (1.9).

Example 1.6. Consider the row matrix R corresponding to the vectors given by its columns and the corresponding null matrix N:

$$R = \begin{bmatrix} 1 & 0 & 2 & 1 \\ 0 & 1 & 1 & 2 \end{bmatrix}, \quad N = \begin{bmatrix} 2 & 1 \\ 1 & 2 \\ -1 & 0 \\ 0 & -1 \end{bmatrix} \tag{1.10}$$

A Kirchhoff graph for (1) this R, (2) any matrix row-equivalent to R, or (3) any set of four vectors with exactly the same dependencies as the columns of R is given in Figure 1.11. Note that the columns of N correspond to a basis of the cycle space for this Kirchhoff graph, while the rows of R correspond to a basis of the cut space.

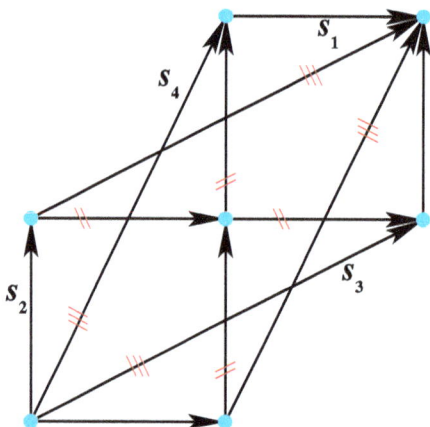

Figure 1.11: A Kirchhoff graph for R in (1.10). Cycles, circuits, or closed walks for this Kirchhoff graph again correspond to vectors spanning Null(R), whereas all the vertex cuts correspond to the vectors in Row(R). Edge vectors here are drawn to match the columns of R. Notice that there are exactly six copies of each edge vector in this Kirchhoff graph.

Example 1.7. Now consider another row matrix corresponding to the vectors given by its columns:

$$R = \begin{bmatrix} 3 & 0 & -2 & 1 \\ 0 & 3 & 1 & -2 \end{bmatrix} \tag{1.11}$$

A Kirchhoff graph for R in (1.11), any matrix row-equivalent to R, or any set of four vectors with exactly the same dependencies as the columns of R is given in Figure 1.12.

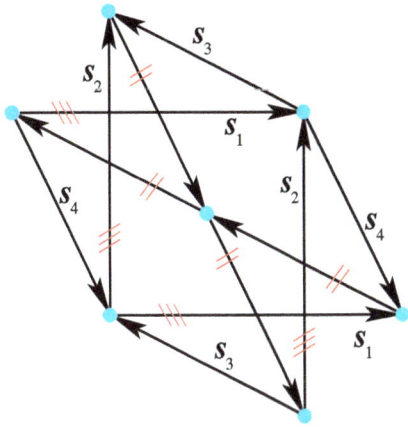

Figure 1.12: A Kirchhoff graph for R in (1.11). Edge vectors again are drawn to match the columns of R, and again there are six copies of each edge vector in this graph.

The row matrices in Example 1.6 and Example 1.7 seem fairly distinct (though of course they are in the general form $[qI|C]$), and the Kirchhoff graphs presented for these two cases look rather different. Perhaps surprisingly, this is a bit misleading: In each case, there are alternate choices for the Kirchhoff graphs that look much more similar. Indeed, for each example, there are in fact four possible distinct Kirchhoff graphs that can be used to generate all known Kirchhoff graphs for these row matrices or their sets of edge vectors. These relationships are discussed in Chapter 4 (Kirchhoff graph construction).

1.4.1 How to recognize a Kirchhoff graph

Up till now, we have always started with a set of vectors or a matrix and then found (somehow) a Kirchhoff graph. What if one starts with a vector graph; how can one decide whether or not such a graph is Kirchhoff? The next two examples should help clarify this situation.

Example 1.8. Consider now the vector graph in Figure 1.13. Although it is not perhaps immediately obvious, this vector graph is also a Kirchhoff graph. To determine this for certain, one needs the matrices R and N corresponding to the vector graph in Figure 1.13. At first glance, one might be tempted to try to find these matrices by viewing each of the vectors in Figure 1.13 as vectors in $\mathbb{R}^2$, determining their components, and viewing these two-component vectors as the columns of a matrix A that is row-equivalent to the row matrix $R = [qI|C]$. Although the graph in Figure 1.13 is certainly a two-dimensional projection of a vector graph, however, there is nothing that says that $\mathbb{R}^2$ is the natural embedding for this vector graph as a Kirchhoff graph.

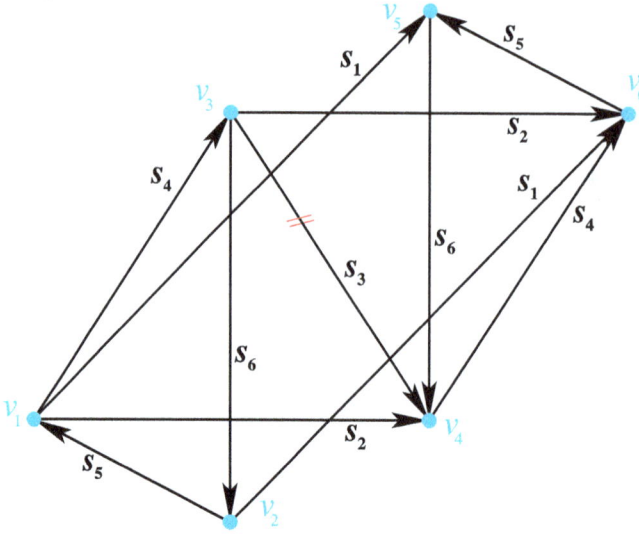

Figure 1.13: A Kirchhoff graph for A in (1.12) and for R in (1.13). Each of the six distinct edge vector appears twice.

To find R and N, form a matrix A whose rows are based on the vertex cuts in the vector graph in Figure 1.13:

$$A = \begin{bmatrix} 1 & 0 & 0 & 0 & 1 & -1 \\ 1 & 1 & 0 & 1 & -1 & 0 \\ 0 & 1 & 2 & -1 & 0 & 1 \end{bmatrix} \qquad (1.12)$$

Here the three rows of A are the vertex cuts for the three left-most vertices v_1, v_2, and v_3 in the vector graph in Figure 1.13; the vertex cuts for the other three vertices are simply the negatives of these first three. Row-reducing A gives R for this vector graph:

$$R = \begin{bmatrix} 1 & 0 & 0 & 0 & 1 & -1 \\ 0 & 1 & 0 & 1 & -2 & 1 \\ 0 & 0 & 1 & -1 & 1 & 0 \end{bmatrix} \qquad (1.13)$$

That R has three rows makes clear that the natural embedding for this vector graph is $\mathbb{R}^3$. In addition, this matrix R implies a null matrix:

$$N = \begin{bmatrix} 0 & 1 & -1 \\ 1 & -2 & 1 \\ -1 & 1 & 0 \\ -1 & 0 & 0 \\ 0 & -1 & 0 \\ 0 & 0 & -1 \end{bmatrix} \qquad (1.14)$$

The columns of this N correspond to a basis for the cycle space. Thus the vector graph shown in Figure 1.13 represents a Kirchhoff graph, and Figure 1.13 is a possible two-dimensional projection of this Kirchhoff graph. In other words, the two-dimensional vector graph given in Figure 1.13 is a projection of a three-dimensional Kirchhoff graph for this R, any matrix row-equivalent to R, or any set of six distinct vectors with exactly the same dependencies as the columns of this R.

Example 1.9. Another example of a vector graph that turns out to be a Kirchhoff graph is shown in Figure 1.14. Again, even though it is drawn in the plane, the plane may not be the nature embedding for this vector graph. Indeed, a careful inspection of the vertices shows that all the vertex cuts are linear combinations of the vertex cuts for the top left three vertices. Thus the incidence matrix for this vector graph is row-equivalent to a smaller matrix

$$A = \begin{bmatrix} 1 & 0 & 1 & 0 & 0 \\ 0 & 0 & -1 & 1 & 1 \\ 0 & 1 & 0 & 1 & -3 \end{bmatrix}$$

By row reducing A into canonical form, one can obtain the R,N-matrix pair for the vector graph in Figure 1.14:

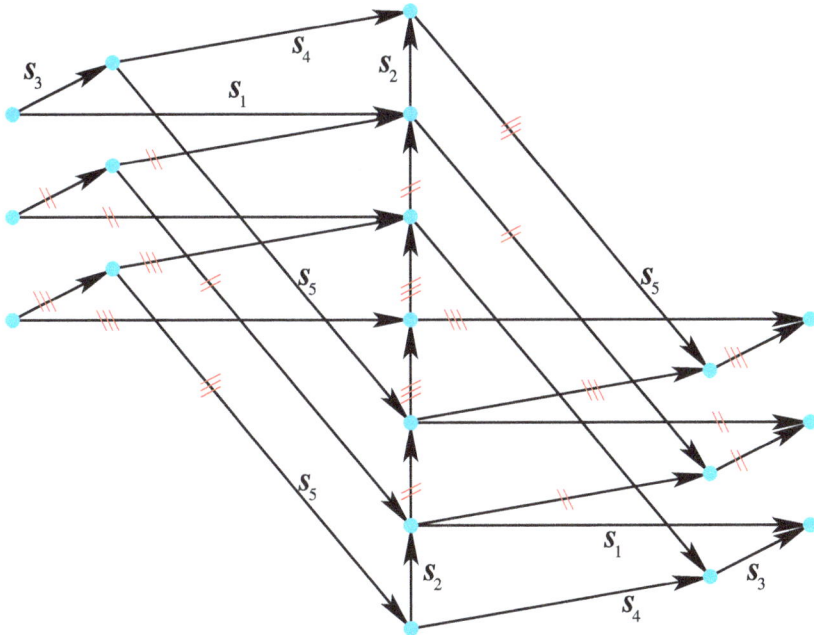

Figure 1.14: A Kirchhoff graph for R in (1.15). Vertices are in blue; red hash marks indicate the number of copies of an edge vector connecting indicated vertices. Cycles or closed walks for this Kirchhoff graph correspond to vectors spanning Null(R), whereas all the vertex cuts correspond to the vectors in Row(R).

$$R = \begin{bmatrix} 1 & 0 & 0 & 1 & 1 \\ 0 & 1 & 0 & 1 & -3 \\ 0 & 0 & 1 & -1 & -1 \end{bmatrix}, \quad N = \begin{bmatrix} 1 & 1 \\ 1 & -3 \\ -1 & -1 \\ -1 & 0 \\ 0 & -1 \end{bmatrix} \tag{1.15}$$

Since the columns of N form a basis for the cycle space of the vector graph in Figure 1.14, this vector graph is a Kirchhoff graph.

One might also wonder if there could be a smaller Kirchhoff graph than that shown in Figure 1.14 for the five vectors in Figure 1.14. Since there are three linearly independent vertex cuts (three linearly independent rows in A and R), there must be at least four vertices in any such Kirchhoff graph. Of course, there are in fact nineteen vertices in the Kirchhoff graph in Figure 1.14. There are also twelve copies of each edge vector in this Kirchhoff graph. Chapter 4 shows, based on the values in R, that twelve is in fact the smallest multiplicity for any Kirchhoff graph for this case. Hence the Kirchhoff graph in Figure 1.14 is close to the very smallest possible Kirchhoff graph for these five edge vectors.

1.5 Overview

One may notice that the Kirchhoff graphs discussed so far have a number of things in common: They are all highly interconnected in a way that will be made clear below. They are also mostly in some sense symmetric unless their edge vectors are distinct. Moreover, they are all uniform in that they all contain the same number of copies of each of their edge vectors. The next two chapters present proofs for results associated with all these observations. Chapter 2 reviews a number of basic Kirchhoff-graph properties. Chapter 3 establishes the uniformity of *most* Kirchhoff graphs. Chapter 4 then discusses numerical approaches to Kirchhoff-graph construction and the implications of these constructions. Specifically, for a given set of vectors S or a given matrix, all known Kirchhoff graphs fall into a family with a certain structure. Chapter 5 considers Kirchhoff graphs over finite fields, and how Kirchhoff graphs and matroids are related. Matroids arise naturally when studying the relationship between graphs and matrices. Although constructing Kirchhoff graphs over the rationals (for rational matrices) can be difficult, the finite nature of $\mathbb{Z}_p$ makes constructing $\mathbb{Z}_p$-Kirchhoff graphs more straightforward. Chapter 6 studies the relationship between equitable edge partitions and Kirchhoff graphs. Chapter 7 looks at the relationship between Maxwell reciprocal diagrams and Kirchhoff graph duals. Finally, Chapter 8 turns back to the origins of Kirchhoff graphs and their implications for chemistry and electrochemistry.

Kirchhoff graphs were developed out of the reaction route graphs of Ravi Datta and Illie Fishtik (see [33] (2010) and its references). They were first defined and discussed by Fehribach [7] (2009); their orthogonality is an extension of the classic result that the cycle space and the cut space of a (standard) graph are orthogonal complements (see, for example, Diestel [5, p. 22] (1997)). Various aspects of Kirchhoff graphs have been studied across the past 15 years or so: Basic properties (Fehribach [8] (2015), Reese [23] (2018)); edge-vector uniformity (Reese et al. [27] (2019), Fehribach [9] (2020)); Kirchhoff graphs over finite fields and matroids (Reese et al. [26] (2018)); Maxwell reciprocal diagrams (Reese et al. [24] (2016)); construction of rank-2, nullity-2 Kirchhoff graphs (Fehribach and McDonald [11] (2019)); equitable edge partitions (Reese et al. [25] (2021)); tilings and Kirchhoff-graph families (Wang and Fehribach [35] (2024)); reaction network circuit diagrams (Fehribach [7] (2009), [8] (2015), [10] (2023)).

2 Basic results

This and the next several chapters build on the mathematical side of the Introduction. For a given set of vectors S, we have already seen that Kirchhoff graphs need not be unique. There are a number of other basic results that follow directly from the definition of Kirchhoff graph or from its association with the row and null matrices, R and N. The first two sections deal with the extreme cases where $k = 0$, $k = 1$, $k = n-1$, or $k = n$. (Recall that n is the number of vectors in S and k is the number of those that are linearly independent.) As it turns out, it is possible to fully characterize Kirchhoff graphs for these cases, though how one views these graphs does depend on how one deals with certain degeneracies—how one sets certain definitions. The three cases discussed in the first section are not of any particular interest, but they are included here for completeness. Except in the first section, it is generally assumed that $1 < k < n$ and that none of the edge vectors is a scalar multiple of any other edge vector. The third section presents a number of basic, general results and observations and is followed by a discussion of a rank theorem for Kirchhoff graphs. This theorem says that at least for the nondegenerate cases, the cut space of any Kirchhoff graph is k-dimensional, while the cycle space is $(n-k)$-dimensional, and there is a basis for $\mathbb{R}^n$ made up of vectors corresponding to vertex cuts and cycles. Finally, a constructive proof is given for the existence of Kirchhoff graphs when the row and null spaces are both two-dimensional.

2.1 Degenerate cases: $k = 0, 1, n$

Recall that at the beginning of the general discussion on Kirchhoff graphs, it was assumed that $1 < k < n$, that is, of the n vectors in our set, at least two are linearly independent, but the entire set is not linearly independent. What does it mean in terms of Kirchhoff graphs if this condition is not satisfied, that is, if $k = 0$, $k = 1$, or $k = n$? In the opinion of the author, these cases are not interesting, but they are mentioned here for completeness.

2.1.1 $k = 0$

When $k = 0$, all the vectors are linearly dependent, implying that all the vectors are the zero vector. For all the edge vectors to be the zero vector, the only possible connected Kirchhoff graph is a single vertex with no edge vectors (or, if one prefers, n zero-length edge vectors); this is a *trivial Kirchhoff graph*. Since R is the zero matrix, N is the identity matrix, and this is consistent with only zero vectors beginning and ending at this single vertex.

https://doi.org/10.1515/9783111408576-002

2.1.2 $k = n$

In this case, S is a set of n linearly independent vectors, R is the identity matrix, and N is the zero matrix. Thus any Kirchhoff graph for these vectors must have no nontrivial cycles. Of a number of possible choices for Kirchhoff graphs in this degenerate case, one choice is a star with n unit vectors, one in each of the n-dimensional directions (see Figure 2.1). An alternative view is to assume that Kirchhoff graphs are *cyclic* vector graphs

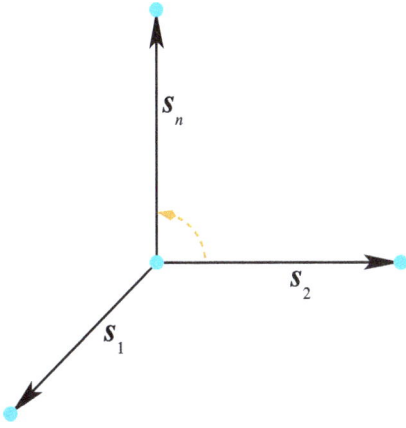

Figure 2.1: Star graph with a single unit vector in each coordinate direction, a possible Kirchhoff graph when R is the identity matrix. The gold arc indicates the vectors in dimensions 3 through $n - 1$.

(see the definition in Section 2.3). This implies that there are no vertices of degree one, in which case the only possible Kirchhoff graph is again the trivial Kirchhoff graph with a single vertex and no edge vectors. Which of these two views to choose is somewhat a matter of taste. In any event, in this case, we are looking for dependencies where there are none.

2.1.3 $k = 1$

In this case, all the vectors are scalar multiples of a single nonzero vector. Assuming that no edge vector is the zero vector, one can construct the Kirchhoff graph in this case as a vector multigraph between two terminal vertices; at these vertices, the vertex cuts are given by the single row of R.

Proposition 2.1.1. *Suppose that R has a single row:*

$$R = [q \quad p_1 \quad p_2 \quad \cdots \quad p_{n-1}]$$

with no zero entries: $q, p_i \neq 0$. Then N has $n-1$ columns, and a Kirchhoff graph for R can be given as a degenerate one-dimensional set of cycles whose edge vectors are scalar multiples of each other.

Proof. In this case, there is the standard null matrix:

$$N = \begin{bmatrix} p_1 & p_2 & p_3 & \cdots & p_{n-2} & p_{n-1} \\ -q & 0 & 0 & \cdots & 0 & 0 \\ 0 & -q & 0 & \cdots & 0 & 0 \\ \vdots & \vdots & \vdots & \vdots & & \vdots \\ 0 & 0 & 0 & \cdots & -q & 0 \\ 0 & 0 & 0 & \cdots & 0 & -q \end{bmatrix}$$

The $n-1$ columns of N represent $n-1$ degenerate cycles. Each of the vertices in the middle of the cycle is a null vertex—a vertex where exactly the same edges enter and exit. On the other hand, the edges entering/exiting the two end vertices satisfy the row space condition, that is, these end vertex cuts are given either by the single row of R or by its negative. □

What happens if some entries of R are zero? The following example deals with this issue.

Example 2.1. Suppose that $R = [1\ -3\ \ 2\ 0]$. Then:

$$N = \begin{bmatrix} -3 & 2 & 0 \\ -1 & 0 & 0 \\ 0 & -1 & 0 \\ 0 & 0 & -1 \end{bmatrix}$$

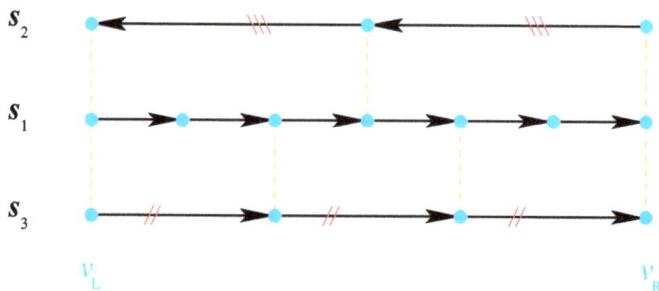

Figure 2.2: A degenerate Kirchhoff graph to illustrate Proposition 2.1.1. The three sets of vectors must be overlaid to form the Kirchhoff graph so that all edge vector sets begin or end at the two end vertices v_L and v_R. The middle vertices in this example all are null vertices where equal numbers of an edge vector both enter and exit. Gold dashed line segments connect identified vertices.

The third column in N is $[0, 0, 0, -1]^T$, so the fourth edge vector forms a cycle by itself, meaning that it begins and ends at the same vertex, and thus $s_4 = \mathbf{0}$. Since R has a single row, the other three edge vectors must be colinear, lying between two vertices v_L and v_R whose vertex cut is given by the row of R. Figure 2.2 shows a degenerate Kirchhoff graph for this example.

2.2 The case $k = n - 1$: a cycle

The next case to consider is definitely not trivial or degenerate; it is the case where the Kirchhoff graph is a single cycle.

Proposition 2.2.1. *Suppose that $k = n - 1 > 1$ and that $N = [p_1 \; p_2 \; \cdots \; p_{n-1} \; -q]^T$ with $q > 0$ and at least two $p_i \neq 0$. Then the row matrix is $(n - 1) \times n$:*

$$R = \begin{bmatrix} q & 0 & 0 & \cdots & 0 & p_1 \\ 0 & q & 0 & \cdots & 0 & p_2 \\ 0 & 0 & q & \cdots & 0 & p_3 \\ \vdots & \vdots & \vdots & \vdots & & \vdots \\ 0 & 0 & 0 & \cdots & q & p_{n-1} \end{bmatrix}$$

A Kirchhoff graph for R can be given as a single cycle with $|p_1| + |p_2| + \cdots + |p_{n-1}| + |q|$ vertices. The edge vectors may appear in the cycle in any order. When edge vectors s_i and s_j are incident on the same vertex, the vertex cut must be a nonzero multiple of $[0, \; \cdots \; p_j \; \cdots \; -p_i \; \cdots, \; 0]$, where p_j is the ith entry, and $-p_i$ is the jth entry. In total, there are $\mathrm{lcm}(q|p_1||p_2| \cdots |p_{n-1}|)$ copies of each edge vector in the graph where the product includes the nonzero p_i. Finally, when $p_i = 0$, the edge vector s_i does not appear in the graph.

Example 2.2. A simple example may be helpful to understand the proposition above: Suppose that $N = [0 \; 1 \; -3 \; 2]^T$. Then the row matrix is 3×4:

$$R = \begin{bmatrix} 2 & 0 & 0 & 0 \\ 0 & 2 & 0 & -1 \\ 0 & 0 & 2 & 3 \end{bmatrix}$$

Figure 2.3 shows a single-cycle Kirchhoff graph for these N and R. Notice that this is *not* the canonical representation for this graph.

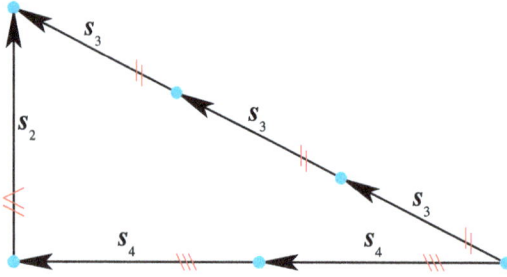

Figure 2.3: A simple cycle to illustrate Proposition 2.2.1. Red hash marks again give the multiplicity of each edge vector (⩽ indicates six copies; it is the Roman numeral VI). The zero in N indicates that the edge vector s_1 is missing, and indeed the graph is naturally embedded in the plane in $\mathbb{R}^3$ defined by s_2 and s_3. The vector representation here is not cannonical.

2.3 General basic results for $1 < k < n$

There are a number of basic observations that are relatively simple to establish but important to keep in mind as we discuss deeper results. The following definitions and propositions give these basic results. Throughout this section and in the chapters to come, assume that $1 < k < n$ and that no edge vector is a scalar multiple of any other edge vector, although occasionally this latter restriction will be relaxed.

We have already noted that any sets of edge vectors that correspond to the same row matrix R or to any matrix A that is row-equivalent to R must have the same set of Kirchhoff graphs. A broader sense of equivalence can actually be helpful when discussing Kirchhoff graphs:

Definition. Sets of edge vectors and their corresponding row matrices are **K-equivalent** if and only if each of the row matrices can be transformed to another by the following operations:

–   a sequence of row operations,
–   a sequence of column permutations,
–   and/or multiplying any column by –1.

In general, any matrix A that can be transformed by these steps to a given row matrix is also K-equivalent to that row matrix.

This definition is justified because along with Kirchhoff graphs being invariant under row operations on their associated matrices, they vary in predictable ways under the other two operations. Specifically if two matrices are the same except that, say, columns i and j are exchanged, then their Kirchhoff graphs are the same except that the ith and jth edge labels are exchanged. Similarly, if two matrices are the same except that the ith columns are the negatives of each other, then their Kirchhoff graphs are the same except that the ith edge vectors are reversed.

The concept of K-equivalence is important of course because if one has a Kirchhoff graph for a given matrix and is looking for a Kirchhoff graph for a K-equivalent matrix, then the desired Kirchhoff graph can be obtained from the existing one.

The next definitions allow us to make exact the term *cyclic* already mentioned.

Definition. Given a vector graph G, its **associated digraph** D (or **associated directed graph**) has the same vertices but has a single directed edge between any two vertices that are connected by one or more copies of an edge vector in G. The direction of each edge in D is the same as the orientation of the corresponding edge vector(s) in the G, written as a function $\psi : G(\mathcal{S}, V) \to D(E, V)$ defined by $\psi : (v_i, s_\ell, v_j, m_{i,j}) \mapsto (v_i, v_j) = e_h$, where $E = \{e_1, e_2, \ldots, e_{n_D}\}$ is the directed edge set having n_D directed edges.

This definition is illustrated by Figure 2.4, which shows a vector graph and its associated digraph. Notice that D is always a *simple* digraph—no loops, no more than one edge connecting a vertex pair.

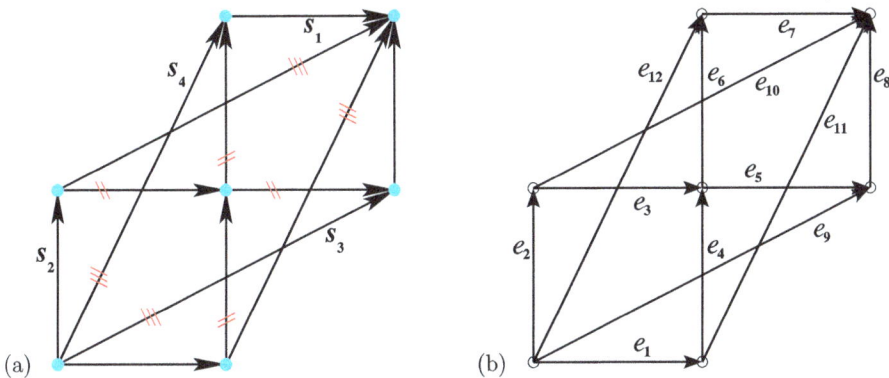

Figure 2.4: A Kirchhoff digraph pair. (a) The Kirchhoff graph G on the left is generated by any set of vectors $\mathcal{S} = \{s_1, s_2, s_3, s_4\}$ where these vectors are represented canonically by the columns of the matrix R in (1.10). Hash marks in the Kirchhoff graph again indicate the number of copies of a given edge vector lying in parallel connecting the same two vertices. (b) The digraph D on the right corresponds to G, having the same vertices as G and a single distinct edge e_j connecting two vertices if and only if there is at least one edge vector connecting those vertices in G.

Definition. A vector graph is **cyclic** if any edge in its associated digraph is contained in a cycle in this digraph, where *cycle* is understood to be a closed walk where only the first and last vertices repeat, and no edges are repeated.

Remarks.

1. Although this requirement could be dropped, our definition of Kirchhoff graphs requires that they are cyclic vector graphs. This requirement rules out the presence of twigs (and thus vertices of degree one) and bridges. These sorts of structures are

excluded because they do not participate in the sorts of vector dependencies that are the central feature of Kirchhoff graphs.

2. In many cases, the cycles of a digraph must traverse each edge in the direction of its orientation. The edges in our graphs, however, correspond to reversible processes (reactions) that have both forward *and* backward directions. Therefore cycles must be allowed to traverse edges regardless of orientation. This view is not unique; it matches the conventions of Bollobás [2] (1998).

Another important property of Kirchhoff graphs is *chirality*:

Definition. Suppose that a vector graph **G** whose natural embedding is the k-dimensional space is projected onto any plane. Its **chiral** is obtained by rotating **G** through 180 degrees about any of its vertices in this plane and then reversing each edge vector.

Although the rotate–reverse process in the previous definition does not precisely produce the mirror image (the meaning of the word "chiral" in chemistry), it is faithful to the key idea. Most importantly here, the process negates each vertex cut and reverses each cycle, leaving the cut space and the cycle space unchanged. All of this implies the following proposition.

Proposition 2.3.1. *The chiral of any Kirchhoff graph is itself a Kirchhoff graph.*

Example 2.3. A simple example of Kirchhoff graphs that are chirals of each other are the triangular Kirchhoff graphs for

$$R = \begin{bmatrix} 1 & 0 & 1 \\ 0 & 1 & 1 \end{bmatrix}$$

shown in Figure 2.5. More complicated examples of chiral pairs are presented in Chapter 4, where computer algorithms will be used to construct Kirchhoff graphs.

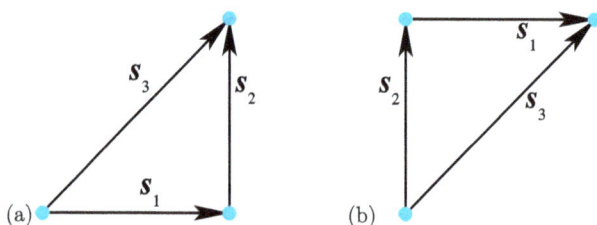

Figure 2.5: Triangular Kirchhoff graphs that are chirals of each other.

The concept of chirality implies a form of symmetry:

Definition. A Kirchhoff graph is **self-chiral** if its chiral is itself. In other words, it is invariant under the rotate–reverse process.

Recalling Example 1.1, both the Square and the Diamond there are self-chiral, like most of the other Kirchhoff graphs discussed in Chapter 1.

2.3.1 Distinct edge vectors

In all the Kirchhoff graph examples considered so far, except for the triangles in several examples, each of the edge vectors appears multiple times in the Kirchhoff graphs. Are there other Kirchhoff graphs where all the edge vectors are distinct? The answer is "Yes", and the following example helps us see why.

Example 2.4. Consider the vector graph shown in Figure 2.6. As in some other examples, it is perhaps not obvious at first that this vector graph is Kirchhoff, but this issue can be decided by considering its vector graph incidence matrix:

$$
B = \begin{bmatrix}
1 & 1 & 1 & 0 & 0 & 0 & 0 & 0 \\
0 & 0 & -1 & -1 & 1 & 0 & 0 & 0 \\
-1 & 0 & 0 & 0 & 0 & 1 & 0 & -1 \\
0 & 0 & 0 & 0 & -1 & -1 & 1 & 0 \\
0 & -1 & 0 & 1 & 0 & 0 & -1 & 1
\end{bmatrix}
$$

Since all the edge vectors are distinct, all but one of the rows of B are linearly independent (since this is an incidence matrix, all the columns sum to zero). One can now verify that each cycle of the vector graph in Figure 2.6 is orthogonal to all the rows of B and therefore to all the vertex cuts of this vector graph. Since this vector graph is also cyclic, it is indeed a Kirchhoff graph. This result can be generalized to any vector graph with distinct edge vectors.

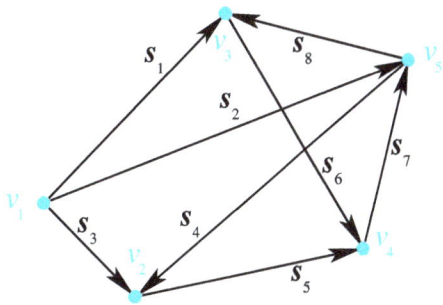

Figure 2.6: A cyclic vector graph where all edge vectors are distinct. Such a vector graph is always Kirchhoff.

The final result of this section basically states that any cyclic vector graph whose edge vectors are all distinct (no edge vector appearing in multiple places in the graph and no multiples copies of edge vectors connecting any two vertices) is in fact a Kirchhoff graph.

Theorem 2.3.1. *Suppose that **G** is a cyclic vector graph having at most one copy of an edge vector connecting any two vertices and no edge vector connecting more than two vertices, so that all the edge vectors are distinct. Then **G** is a Kirchhoff graph for its set of edge vectors, its incidence matrix, and any matrix A that is row-equivalent to this incidence matrix.*

Proof. In this case, the vector graph **G** is essentially a standard cyclic digraph, that is, the digraph constructed by identifying each distinct edge vector with a directed edge having the same direction as the orientation of the edge vector. Thus this proposition for vector graphs is just the standard orthogonality result for digraphs (see, for example, Diestel [5, p. 22] (1997) or Bollobás [2, p. 53] (1998)). □

Remark. If a matrix is the incidence matrix for a Kirchhoff graph whose all edge vectors are distinct or for a simple cyclic digraph, then:
(1) All entries are 0 or ±1.
(2) Each column has exactly one 1 and one –1.
(3) No two columns have both their nonzero entries in the same rows.
(4) Each row has at least two nonzero entries.

The previous example (Example 2.4) might make us think that every cyclic vector graph is a Kirchhoff graph for some row matrix R—that the row-space, cycle-space orthogonality condition is redundant. This is not the case where edge vectors appear multiple times, as the following counterexample shows.

Example 2.5. Consider the leftmost vector graph in Figure 2.7. If this graph is a Kirchhoff graph for some matrix A, then $[1, 1, -1]^T$ must be in Null(A), and $[1, 1, 1]$ must be in Row(A), which of course is impossible since these vectors are not orthogonal. So the cy-

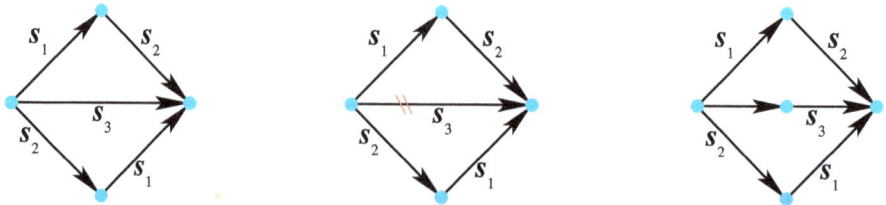

Figure 2.7: Three similar vector graphs: the leftmost is not the Kirchhoff graph of any matrix; the center and rightmost are Kirchhoff graphs. Note that the Kirchhoff graph in the center is not minimal; it is simply the union of two triangular Kirchhoff graphs, each triangular graph having one copy of each edge vector.

cle and cut spaces for this graph are not orthogonal. There are two possible Kirchhoff graphs similar to this non-Kirchhoff graph, and they are shown in the center and on the right in Figure 2.7.

2.4 Cut space, cycle space orthogonal complementarity

By the definition there must be a bijection between the cycles or closed walks in any Kirchhoff graph and integral elements of $\text{Null}(R)$. As a consequence of the existence of this bijection, the columns of N correspond to a basis for the cycle space of that Kirchhoff graph. In addition, the second condition of the definition requires that all the vertex cuts in any Kirchhoff graph must correspond to integral elements of $\text{Row}(R)$. The reader may have noticed that at least in the nontrivial cases, not only do these vertex cuts lie in this row space, but they also span it. The next theorem shows that this is indeed the case, at least for Kirchhoff graphs over the rationals.

Theorem 2.4.1. *Suppose that G is a Kirchhoff graph for a certain set of n edge vectors S and thus for a certain row-null matrix pair R and N. Suppose also that all the edge vectors in S are required to represent the cycles of G and that there are k linearly independent vectors in S with $1 < k < n$. Then the rows of R correspond to a basis for the cut space of G.*

Proof. Recall that by the definition of Kirchhoff graph, the vertex cuts of G lie in $\text{Row}(R)$. Suppose that all the vertex cuts are linear combinations of a proper subset of the rows of R, that is, that there exists at least one row of R not needed to represent the vertex cuts of G. Say that the ith row for $1 \le i \le k$ is not needed for any vertex cut. Then s_i is not present in any vertex cut of G since the ith column entry in each row of R is zero, except for the ith row. Thus s_i is not required for any cycle or closed walk in G, contradicting that all edge vectors of S appear in some cycle for G. $\square$

Remark. We have already had examples of degenerate Kirchhoff graphs where certain vectors from S are not present in the Kirchhoff-graph cycles and where Theorem 2.4.1 does not hold.

A second result of this type is that if a vector graph for a set of edge vectors has a full set of vertex cuts, then that vector graph is a Kirchhoff graph.

Theorem 2.4.2. *Suppose that G is a cyclic vector graph for a certain set of edge vectors S and thus for a certain row-null matrix pair R and N. If the vertex cuts of G correspond to vectors that span $\text{Row}(R)$ and thus all edge vectors of S occur in G, then each column of N and thus all elements of $\text{Null}(R)$ correspond to a cycle, circuit, or closed walk in G. Also, all the cycles, circuits, and closed walks of G correspond to elements of $\text{Null}(R)$. Thus G is a Kirchhoff graph.*

Proof. First, suppose that the vertex cuts of G correspond to row vectors that span
Row(R). If there is a column of N that does not correspond to a cycle or closed walk
in G (say, it is the ith column of N), then no element of the span of the other $n - k - 1$
columns of N contains a nonzero entry for the $(k+i)$th edge vector. Thus this edge vector
cannot appear in G, contradicting that all edge vectors appear in G.

On the other hand, if a cycle, circuit, or closed walk in G does not correspond to an
element of Null(R), then this would imply a linear combination of the edge vectors in S
that was not present in the original definition of R and N, a contradiction. $\square$

Remark. Theorem 2.4.2 has an important consequence for constructing Kirchhoff
graphs: If one constructs a cyclic vector graph so that the vertex cuts of G correspond
to row vectors that span Row(R) and it includes the full set of edge vectors, then this
vector graph is in fact a Kirchhoff graph. This result is used implicitly in both of the
construction algorithms in Chapter 4.

2.5 Rank-two, nullity-two Kirchhoff graphs

Beyond the most basic cases $k = 0, 1, n - 1, n$, there is another general case where Kirchhoff graphs can directly be constructed: when the matrix R has rank two and nullity two,
that is, $n = 4$ and $k = 2$ (and thus $n - k = 2$). This case was studied by Fehribach and
McDonald [11] (2018). One reason why this construction is possible is that essentially just
two cases need to be considered. Trying to make this sort of construction for larger row
matrices becomes unwieldy; other computational approaches are superior.

Theorem 2.5.1. *Every $m \times n$ rational matrix A has a Kirchhoff graph if $n \le 4$.*

All cases here *except* for $n = 4$, $k = 2$ have already been dealt with. Although row
equivalence is the fundamental equivalence for matrices in the construction of Kirchhoff graphs, the somewhat broader sense of K-equivalence is more useful here since
Kirchhoff graphs change in predictable ways under this equivalence: either edge vectors are relabeled, or at least one vector is reversed.

Lemma 2.5.1. *Every rank-two, nullity-two matrix over the rationals with no column a
multiple of another is K-equivalent to a specific row matrix:*

$$R = \begin{bmatrix} q & 0 & p & r \\ 0 & q & t & u \end{bmatrix} \tag{2.1}$$

where q, p, t, and u are all positive integers, and r is either a positive or negative integer.

Proof. Every rank-two, nullity-two matrix must have four columns and two pivots. By
K-equivalence the first two columns can be made the pivot columns with unit pivots.
Each row can be multiplied by q, the least common multiple of the denominators of the
K-equivalent matrix with unit pivots. Then both pivots are q, and all the entries in the

nonpivot columns are nonzero integers (if any of p, t, r, or u were zero, then some edge vector would be a scalar multiple of another). Now if any of the nonzero entries other than r are negative, then they may all be made positive through a sequence of row and column multiplications by –1. Thus all the nonzero entries of R can be positive, except for r.   □

Remark. Nothing is lost by avoiding the case where one column of the matrix is a multiple of another since this case degenerates to a Kirchhoff graph with three or fewer edge vectors, implying that either the rank or the nullity is zero or one, and again these cases have already been resolved. So by the lemma there are only two cases that need to be considered: first where $r < 0$ and second where $r > 0$.

2.5.1 The $r < 0$ construction

Given the row matrix R defined in Lemma 2.5.1, the corresponding null matrix can be written as before:

$$N = \begin{bmatrix} p & r \\ t & u \\ -q & 0 \\ 0 & -q \end{bmatrix} \tag{2.2}$$

To construct a Kirchhoff graph for R and any row-equivalent matrix, one must find a vector graph whose vertex cuts span the row space of R and whose cycles span the null space of R. The columns of R provide a convenient representation of the edge vectors s_1, s_2, s_3, and s_4 that are displayed in Figure 2.8.

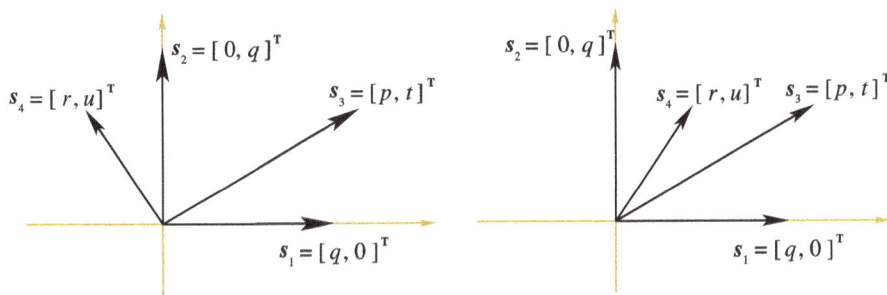

Figure 2.8: Edge vectors for the rank-two, nullity-two construction based on the columns of R. All the nonzero entries are positive, except that $r < 0$ on the left and $r > 0$ on the right.

Our actual construction begins by forming a *Kirchhoff block*, a rectangle or parallelogram based on one cycle of N that can tile the plane. Such a Kirchhoff block for the

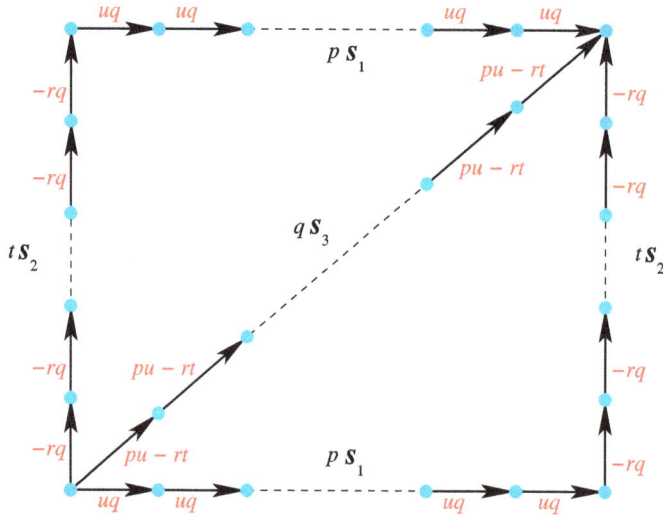

Figure 2.9: The edge-weighted Kirchhoff block for R in (2.1) when $r < 0$. There are $p\mathbf{s}_1$ in sequence along the bottom and top, $t\mathbf{s}_2$ along the vertical sides, and $q\mathbf{s}_3$ on the diagonal. The cycle corresponds to the first column of N in (2.2): $p\mathbf{s}_1 + t\mathbf{s}_2 - q\mathbf{s}_3 = \mathbf{0}$. The multiplicities (in red) for the lower-left and upper-right corner vertices correspond to a linear combination of the rows of R: these vertex cuts are $\pm[uq, -rq, pu - rt, 0]$. Recall that here $-r > 0$, so all vectors have the same orientation at these two corners. The lower-right and upper-left corner vertices do not yet correspond to the row space of R. All other vertices are null vertices; the same edge vectors enter and exit.

current case is shown in Figure 2.9. Here the nontrivial cycle corresponds to the first column of N. In addition, the multiplicities (weights, valences, or numbers) of edge vectors into or out of the lower-left and upper-right vertices are assigned so that the vertex cuts are the same linear combination (up to sign) of the rows of R. These two corner vertices thus lie in the row space of R, though the remaining two corner vertices do not. All other vertices are null vertices.

The next step in our construction is to tile an array of blocks together. The key vertices are those on the outer array edges where the blocks meet. Each new block is an exact copy of the initial block in Figure 2.9, *except* that the multiplicities for the outer edge vectors are multiplied by the level of the block. This arrangement is shown in Figure 2.10 for the lower-left corner of the array. Along the left side at vertex v_1 (the top vertex of the lower block), $-rq$ copies of $\mathbf{s}_2$ enter, whereas $-2rq$ copies of $\mathbf{s}_2$ exit. As a result, this vertex has the same vertex cut as the lower-left-most vertex in the array, v_0. Similarly for v_2, the right-most vertex of the left block along the bottom of the array; here uq copies of $\mathbf{s}_1$ enter and $2uq$ copies exit, implying again that this vertex has the same vertex cut as the lower left-most vertex. This tiling continues with pu layers of blocks from the bottom to the top of the array and with $-rt$ columns of blocks moving from left to right. At this point in the construction, all vertices are either null vertices with the same vectors having the same multiplicities both entering and exiting or are

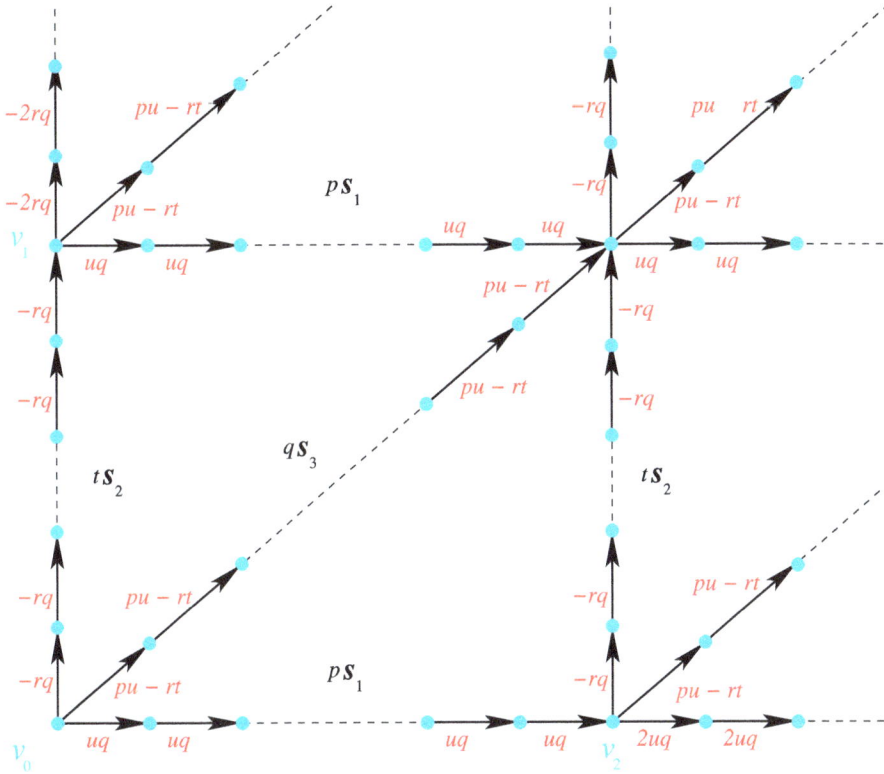

Figure 2.10: Lower-left corner of an array of Kirchhoff blocks. The vertex cut for the vertex v_1 on the left array edge where the original Kirchhoff block meets the left-most block of the second layer is $[uq, -2rq - (-rq), pu - rt, 0] = [uq, -rq, pu - rt, 0]$, the same as for v_0 at the lower-left corner of the original Kirchhoff block. In addition, the vertex cut for the vertex v_2 on the lower array edge where the original Kirchhoff block meets the lowest block of the second column is $[2uq - uq, -rq, pu - rt, 0] = [uq, -rq, pu - rt, 0]$. All other vertices in the diagram are null vertices in that the same edge vectors enter and exit.

at the lower left or upper right of some Kirchhoff block, *except* for two, the one at the upper-left corner of the array and the one at the lower-right corner.

These final two vertices constitute loose ends that must now be tied up by finally introducing the fourth edge vector, s_4. This final step in the construction is shown in Figure 2.11. A sequence of s_4 edge vectors exits the lower-right corner vertex of the array, heading for the upper-left corner vertex. Recalling the second column of N in (2.2), we have that $rt(ps_1) + pu(ts_2) = pt(qs_4)$, meaning that a sequence of ptq of the fourth edge vector will move from the lower-right array corner vertex to the upper-left array corner vertex. If there are $-ru(pu - rt)$ copies of each of these s_4 vectors, that is, the multiplicity of each edge vector is $-ru(pu - rt)$, then both of these final two vertices lie in the row space of R. With the introduction of the fourth edge vector, the construction of the rank-two, nullity-two Kirchhoff graph is complete; the full graph is shown in Figure 2.12.

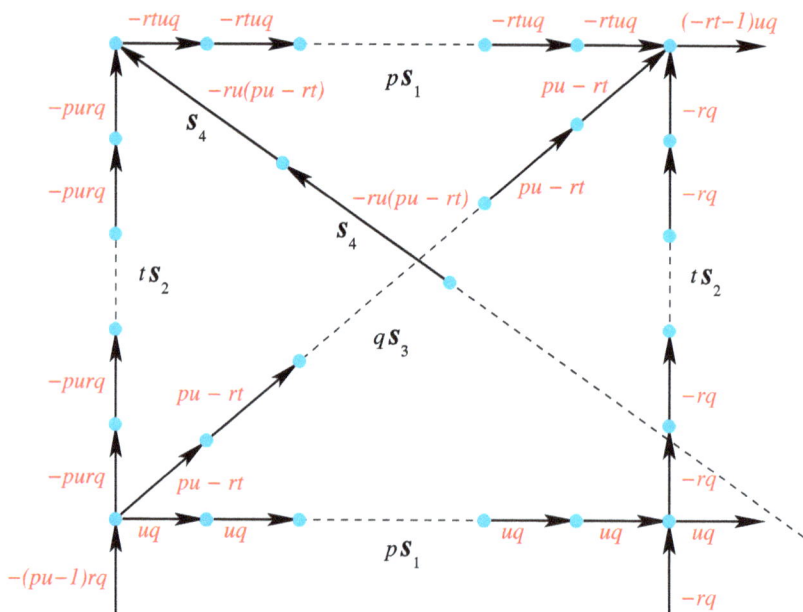

Figure 2.11: The upper-left corner block in the array. Here $-ru(pu-rt)$ copies of s_4 enter the upper-left corner vertex to make its vertex cut $-ru[tq, -pq, 0, -(pu-rt)]$ and thereby place this vertex in the row space of R.

2.5.2 The $r > 0$ construction

The construction when $r > 0$ is similar to that for $r < 0$, except that different edge vectors must be used for the Kirchhoff block so that the remaining edge vector can be used to tie up the loose ends. Without loss of generality, assume that $pu - rt > 0,$[1] as shown in Figure 2.8. Consider the Kirchhoff block formed using s_1, s_3, and s_4, as shown in Figure 2.13. This Kirchhoff block is then tiled as before into an array, with $rt(pu-rt)s_1$ edge vectors in sequence horizontally across the base and $(pu-rt)tqs_4$ edge vectors diagonally upward along the left side of the array. Again the loose ends must be tied up: this is done with $(pu-rt)tus_2$ edge vectors in sequence vertically from the lower-right corner of the array to the upper-left corner. Thus $rt(pu-rt)s_1 + (pu-rt)tus_2 = (pu-rt)tqs_4$. The multiplicities for the array are extended from the multiplicities of the block as in the $r < 0$ case, making this array a Kirchhoff graph.

Remarks.
1. Depending on the exact values of the entries q, p, r, t, and u, it may be possible to find a smaller Kirchhoff graph than that constructed above. For example, if

---

1 If $pu - rt < 0$, then switch the labeling of s_3 and s_4 since $p, r, t, u > 0$.

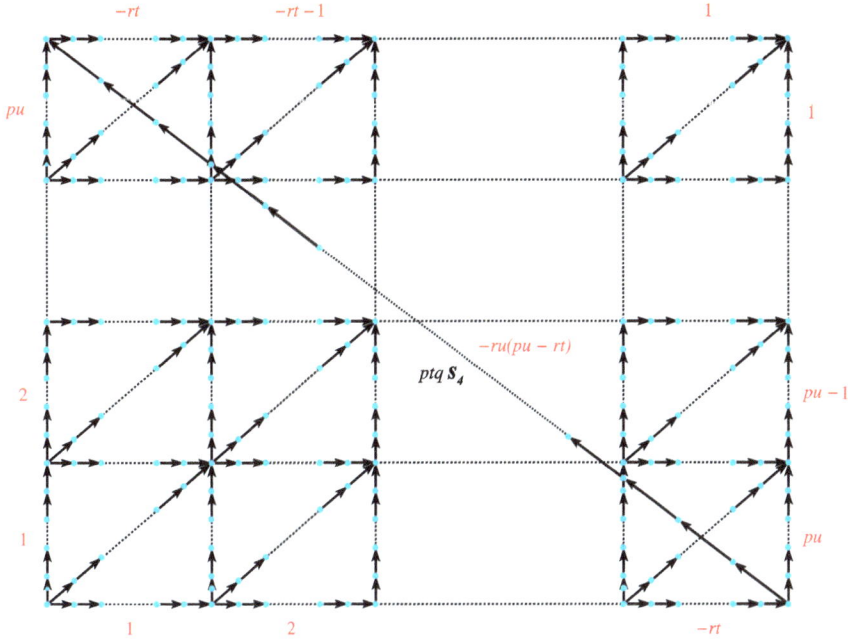

Figure 2.12: Full Kirchhoff graph for this rank-two, nullity-two construction. The full graph is a $-rt \times pu$ array of Kirchhoff blocks with s_4 in sequence ptq times from the lower right corner to the upper left corner. There are then $-ru(pu - rt)$ copies of s_4 connecting in parallel each pair of adjacent vertices along this diagonal. This diagonal of s_4 vectors places the final two vertices of the array into the row space of R. In total, there are $(-rt)puq(pu - rt)$ copies of each vector in this Kirchhoff graph.

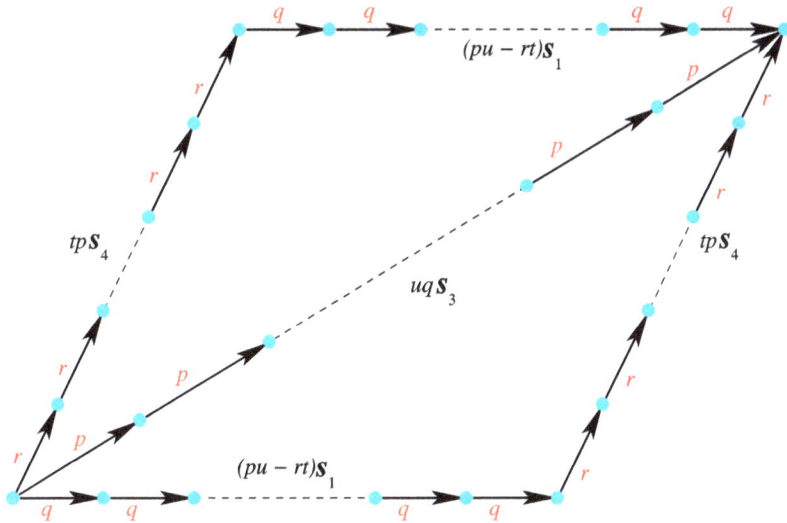

Figure 2.13: A Kirchhoff block when $r > 0$. Taking the difference of the two columns in N, we have that $(pu - rt)s_1 + tqs_4 = uqs_3$. Also the vertex cuts for the lower-left and upper-right vertices are $\pm[q, 0, p, r]$, which is the second row of R.

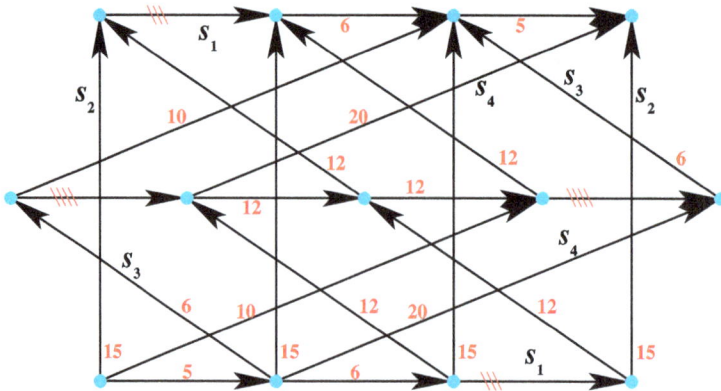

Figure 2.14: Relatively small Kirchhoff graph for a specific R compared to the general one constructed above. Notice that the horizontal and vertical vectors in this case are drawn on different scales to allow room for the edge labels. In this case, hash marks indicate when three or four copies of a vector connect adjacent vertices, whereas red Arabic numerals give this number when there are five or more. There are sixty copies of each edge vector in this Kirchhoff graph.

$$R = \begin{bmatrix} 2 & 0 & 5 & -3 \\ 0 & 2 & 1 & 1 \end{bmatrix}$$

then the Kirchhoff graph constructed above has 240 copies of each of its edge vectors. Figure 2.14 shows a smaller Kirchhoff graph for this R and its column vectors; this alternative Kirchhoff graph has only sixty copies of each edge vector. Indeed this seems to be among the smallest possible Kirchhoff graphs for this R (see Section 4.3.2). Only when q, p, u, r, t, and $pu - rt$ are all relatively prime will the Kirchhoff graph constructed here likely be as small as possible. Smaller Kirchhoff graphs that more clearly show the vector dependencies are in general better.

2. As was mentioned above, a similar construction should be possible for higher-dimensional cases—rank-three, nullity-two or rank-two, nullity-three, for example. Row matrices of these sizes, however, seem to cause a much larger number of cases, not just two, and this makes such a construction much more difficult.

3 Kirchhoff graph uniformity

Many readers likely have noticed that in the examples of Kirchhoff graphs shown so far, each edge vector appears the same number of times. So, for example, in Example 1.1, each edge vector appears twice. One might wonder if this is always the case for Kirchhoff graphs. The short answer is almost, but not quite; the full answer is the goal of this chapter.

Definition. For a vector graph G with edge vectors $\{s_1, s_2, \ldots, s_n\}$, let m_i be the **multiplicity** of the edge vector s_i in G, that is, the number of times that s_i occurs in G. A Kirchhoff graph is **uniform** if each edge vector appears the same number of times: there is a multiplicity $m \in \mathbb{Z}^+$ such that $m_i = m$ for all $1 \le i \le n$.

All the examples of Kirchhoff graphs presented so far have been uniform; the next example is *not*.

Example 3.1. Consider the edge vectors

$$s_1 = \begin{bmatrix} 1 \\ 0 \\ 0 \end{bmatrix}, \quad s_2 = \begin{bmatrix} 0 \\ 1 \\ 0 \end{bmatrix}, \quad s_3 = \begin{bmatrix} 0 \\ 0 \\ 2 \end{bmatrix}, \quad s_4 = \begin{bmatrix} 1 \\ -1 \\ 0 \end{bmatrix}$$

with corresponding canonical row and null matrices:

$$R = \begin{bmatrix} 1 & 0 & 0 & 1 \\ 0 & 1 & 0 & -1 \\ 0 & 0 & 1 & 0 \end{bmatrix}, \quad N = \begin{bmatrix} 1 \\ -1 \\ 0 \\ -1 \end{bmatrix} \tag{3.1}$$

A possible nonuniform Kirchhoff graph for these four edge vectors is shown in Figure 3.1. Notice that this Kirchhoff graph has two copies of edge vectors s_1, s_2, and s_4 but three copies of s_3. Also notice that this third edge vector corresponds to a row and column in R whose entries are all 0 except for the single 1 in the $(3, 3)$ position. The zero in the third entry of N means that the vector s_3 cannot participate in any nontrivial cycle in a Kirchhoff graph for these edge vectors, and this nonparticipation is reflected in this example. Looked at another way, this zero implies that there is no control over the number of copies of s_3 in the graph. All this is key to why Kirchhoff graphs for this set of edge vectors need not be uniform. The exact condition that guarantees that a Kirchhoff graph be uniform is *vector 2-connectedness* (see Theorem 3.2.1), but before turning our attention directly to vector 2-connectedness, let us develop a linear-algebraic characterization that is equivalent to a vector graph being Kirchhoff.

https://doi.org/10.1515/9783111408576-003

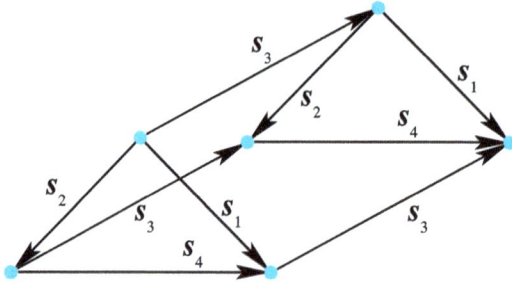

Figure 3.1: Tent Kirchhoff graph. Notice that this Kirchhoff graph is *not* uniform: there are two copies of s_1, s_2, and s_4 but three copies of s_3.

3.1 Linear-algebraic characterization

Finding a linear-algebraic characterization that is equivalent to a vector graph being Kirchhoff begins with constructing an *associated digraph* based on the Kirchhoff graph: Given a vector graph G (perhaps Kirchhoff, perhaps not), define a digraph D having the same vertex set V, but having a single directed edge between any two vertices that are connected by at least one edge vector. Let the digraph edge direction be the same as the positive orientation of the corresponding edge vector in the vector graph. This is the same digraph used in the definition of *cyclic* above. As before, mathematically this means that there is a function $\psi : G(S, V) \rightarrow D(E, V)$ defined by $\psi : (v_i, s_\ell, v_j, m_{ij}) \mapsto (v_i, v_j) = e_h$, where $E = \{e_1, e_2, \dots, e_{n_D}\}$ is the directed edge set having n_D directed edges. An example of a vector graph and its associated digraph is again shown in Figure 3.2.

Notice that although G may have multiple copies of a given vector between two given vertices, D is always *simple*: there is at most one directed edge between any two vertices in D. Also notice that although the same edge vector may be associated with many vertices in G, each edge in D is associated with exactly two vertices since D is a standard digraph. Finally notice that all cycles, circuits, or closed walks in G correspond to cycles, circuits, or closed walks in D.

Example 3.2. As an example of a vector graph (that is indeed also a Kirchhoff graph) and its corresponding digraph, consider the pair shown in Figure 3.2. In this example,

$$R = \begin{bmatrix} 1 & 0 & 2 & 1 \\ 0 & 1 & 1 & 2 \end{bmatrix}, \quad N = \begin{bmatrix} 2 & 1 \\ 1 & 2 \\ -1 & 0 \\ 0 & -1 \end{bmatrix} \tag{3.2}$$

and the vector edges in Figure 3.2 are drawn based on the columns of R.

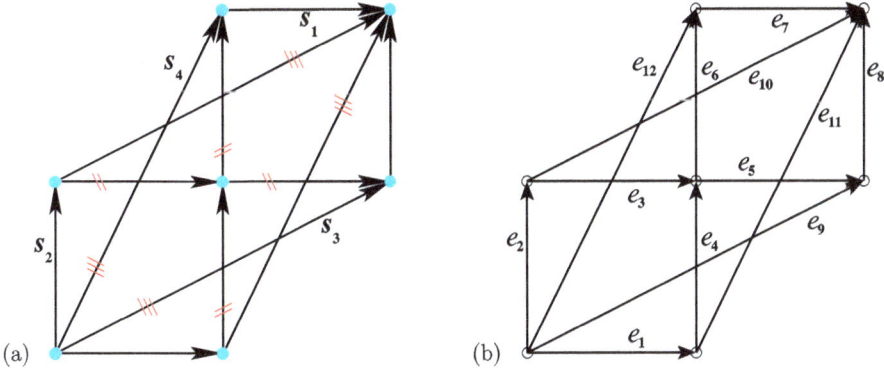

Figure 3.2: A Kirchhoff digraph pair. (a) The Kirchhoff graph **G** on the left is generated by any set of vectors $S = \{s_1, s_2, s_3, s_4\}$, where these vectors are represented canonically by the columns of the matrix R in (1.10) and (3.2). (b) The digraph D on the right corresponds to **G**, having the same vertices as **G** and a single, distinct edge e_j connecting two vertices if and only if there is at least one edge vector connecting those vertices in **G**.

Definition. Let n_D be the number of directed edges in D. Let $S = [s_{ij}]$ be the **assignment matrix** for a given vector graph G: the ith directed edge e_i in the digraph D is assigned s_{ij} copies of the jth edge vector s_j in G when e_i and s_j connect the same two vertices in V. Let $T = [t_{ij}]$ be the **characteristic matrix**: $t_{ij} = 1$ if the ith directed edge in D and the jth edge vector in G connect the same vertices; $t_{ij} = 0$ otherwise. So S and T are both $n_D \times n$ matrices, and T has a 0 whenever S has a 0 entry, and 1 as an entry whenever S has a positive entry. Let B be the **incidence matrix** for D: the ijth entry of B is 1 if e_j exits vertex v_i, -1 if e_j enters vertex v_i, and 0 otherwise.

Remark. With these definitions, BS is the **incidence matrix** for **G**.

Example 3.3. Return to our example:

$$
B = \begin{array}{c} \\ v_1 \\ v_2 \\ v_3 \\ v_4 \\ v_5 \\ v_6 \\ v_7 \end{array}
\begin{array}{cccccccccccc}
e_1 & e_2 & e_3 & e_4 & e_5 & e_6 & e_7 & e_8 & e_9 & e_{10} & e_{11} & e_{12} \\
\left[\begin{array}{cccccccccccc}
1 & 1 & 0 & 0 & 0 & 0 & 0 & 0 & 1 & 0 & 0 & 1 \\
-1 & 0 & 0 & 1 & 0 & 0 & 0 & 0 & 0 & 0 & 1 & 0 \\
0 & -1 & 1 & 0 & 0 & 0 & 0 & 0 & 0 & 1 & 0 & 0 \\
0 & 0 & -1 & -1 & 1 & 1 & 0 & 0 & 0 & 0 & 0 & 0 \\
0 & 0 & 0 & 0 & -1 & 0 & 0 & 1 & -1 & 0 & 0 & 0 \\
0 & 0 & 0 & 0 & 0 & -1 & 1 & 0 & 0 & 0 & 0 & -1 \\
0 & 0 & 0 & 0 & 0 & 0 & -1 & -1 & 0 & -1 & -1 & 0
\end{array}\right]
\end{array}
$$

$$S = \begin{array}{c} \\ e_1 \\ e_2 \\ e_3 \\ e_4 \\ e_5 \\ e_6 \\ e_7 \\ e_8 \\ e_9 \\ e_{10} \\ e_{11} \\ e_{12} \end{array} \begin{array}{cccc} s_1 & s_2 & s_3 & s_4 \\ \left[\begin{array}{cccc} 1 & 0 & 0 & 0 \\ 0 & 1 & 0 & 0 \\ 2 & 0 & 0 & 0 \\ 0 & 2 & 0 & 0 \\ 2 & 0 & 0 & 0 \\ 0 & 2 & 0 & 0 \\ 1 & 0 & 0 & 0 \\ 0 & 1 & 0 & 0 \\ 0 & 0 & 3 & 0 \\ 0 & 0 & 3 & 0 \\ 0 & 0 & 0 & 3 \\ 0 & 0 & 0 & 3 \end{array}\right] \end{array} \quad \text{and} \quad BS = \begin{array}{c} \\ v_1 \\ v_2 \\ v_3 \\ v_4 \\ v_5 \\ v_6 \\ v_7 \end{array} \begin{array}{cccc} s_1 & s_2 & s_3 & s_4 \\ \left[\begin{array}{cccc} 1 & 1 & 3 & 3 \\ -1 & 2 & 0 & 3 \\ 2 & -1 & 3 & 0 \\ 0 & 0 & 0 & 0 \\ -2 & 1 & -3 & 0 \\ 1 & -2 & 0 & -3 \\ -1 & -1 & -3 & -3 \end{array}\right] \end{array}$$

The matrix T is the same as S except that each 1, 2, or 3 in S is replaced by a 1 in T.

The following theorem gives the linear-algebric characterization needed to prove the uniformity of a Kirchhoff graph. It is similar to Theorem 1 of Reese et al. [27] (2019) but involves smaller matrices.

Theorem 3.1.1. *A vector graph G is Kirchhoff if and only if:*

$$T^{\mathrm{T}} \operatorname{Null}(B) = \operatorname{Null}(BS). \tag{3.3}$$

Proof. Because BS is the incidence matrix for the vector graph, there is automatically a cycle basis for G corresponding to a basis for $\operatorname{Null}(BS)$. Given a vertex $v \in V(G) = V(D)$, let $\lambda(v)\,[\lambda(v)]$ be the vertex cut or incidence vector for v in $G\,[D]$. Similarly, if $c \in C(G) \cong C(D)$, then let $\chi(c)\,[\chi(c)]$ be the cycle vector for c in $G\,[D]$.

($\Leftarrow$) Suppose that (3.3) holds. Let $v \in V(G) = V(D)$ and $c \in C(G) = C(D)$. Then[1]
- $\lambda(v) \in \operatorname{row}(B) \Rightarrow \lambda(v) = \lambda(v)S \in \operatorname{Row}(BS)$;
- $(\chi(c))^{\mathrm{T}} \in \operatorname{Null}(B) \Rightarrow (\chi(c))^{\mathrm{T}} = (\chi(c)T)^{\mathrm{T}} = T^{\mathrm{T}}(\chi(c))^{\mathrm{T}} \in T^{\mathrm{T}} \operatorname{Null}(B) = \operatorname{Null}(BS)$.

Thus $\lambda(v) \perp \chi(c)$ for all v, c.

($\Rightarrow$) Now suppose $\lambda(v)(\chi(c))^{\mathrm{T}} = 0$ for all $v \in V(G)$ and $c \in C(G)$. Since G is Kirchhoff, $(\chi(c))^{\mathrm{T}} \in \operatorname{Null}(BS)$. At the same time, $(\chi(c))^{\mathrm{T}} = (\chi(c)T)^{\mathrm{T}} = T^{\mathrm{T}}(\chi(c))^{\mathrm{T}} \in T^{\mathrm{T}} \operatorname{Null}(B)$. So vectors are in $\operatorname{Null}(BS)$ if and only if they are also in $T^{\mathrm{T}} \operatorname{Null}(B)$. $\square$

Our linear-algebraic characterization of a Kirchhoff graph can now be used to prove a lemma that is a key to showing when Kirchhoff graphs are uniform.

---

1 Row(X) is the row space of a matrix X; row(X) is the set of rows of X.

Definition. Let $H := T^T S$; so H is a diagonal matrix with $H_{jj} = m_j$, the number of times the vector s_j appears in G.

Lemma 3.1.1. *If G is Kirchhoff, then $v \in \text{Null}(BS)$ implies $Hv \in \text{Null}(BS)$.*

Proof. $v \in \text{Null}(BS) \Rightarrow BSv = 0 \Rightarrow Sv \in \text{Null}(B) \Rightarrow Hv = T^T Sv \in T^T \text{Null}(B) = \text{Null}(BS)$. □

The next theorem deals with the case where the null space is one-dimensional and the Kirchhoff graph is a single cycle. This result follows from Proposition 2.3 in Fehribach [8] (2015) and was discussed in detail by Reese et al. [27] (2019). To deal with more-general Kirchhoff graphs such as that in Figure 3.2, the next section introduces the concept of 2-connectedness for a vector graph.

Theorem 3.1.2. *Suppose that $\text{Null}(BS) = \text{span}(b)$ with $b = [b_1, b_2, \ldots b_n]^T$. If $b_j \neq 0$, then $m_j = m$, independent of j. Hence each edge vector appears m times, provided that the corresponding entry in b is nonzero.*

Proof. Since $b \in \text{Null}(BS)$, by the previous lemma, $Hb \in \text{Null}(BS)$. Since $\text{Null}(BS)$ is spanned by a single vector b, $Hb = mb$ for some m. Then m is an eigenvalue for the diagonal matrix H, implying that $m_j b_j = m b_j$. Thus $m_j = m$, provided that $b_j \neq 0$. □

3.2 Vector 2-connectedness

The previous theorem can be generalized to other Kirchhoff graphs when there are non-trivial cycles that weave their way through the vector graph. The exact sense of *nontrivial* needed here is *vector 2-connected*:

Definition. A vector graph G is **vector 2-connected** if for any pair of vector edges s_i and s_j, there exists a cycle c such that the cycle vector $\chi(c)$ is nonzero with respect to both s_i and s_j.

This definition and its implications were first discussed by Reese et al. [27] (2019). The name for this concept (vector 2-connected) comes from the fact that one must remove *all* copies of at least two edge vectors to disconnect a vector 2-connected vector graph.

Theorem 3.2.1. *Every vector 2-connected Kirchhoff graph is uniform.*

Proof. First suppose that the upper block C in the matrix N has no zero entries. Let b_j denote the jth column of N; by Lemma 3.1.1, $b_j \in \text{Null}(BS)$ implies

$$Hb_j = \gamma_1 b_1 + \gamma_2 b_2 + \cdots + \gamma_{n-k} b_{n-k}$$

for some $\gamma_j \in \mathbb{Q}$. Since H is a diagonal matrix, and because of the lower identity block in N, one can go entry by entry through b_j to find that $\gamma_i = 0$ for $i \neq j$, implying that

$$Hb_j = \gamma_j b_j$$

for all $1 \leq j \leq n - k$, which makes γ_j an eigenvalue of H, and unless the ith entry in b_j is zero, all the corresponding entries on the diagonal of H must be the same. This must be true for each j. Again, because of the block structure of N and the lack of zeros in C, all the entries on the diagonal of H must be the same. Hence $H = mI$.

Now suppose that the upper block C has some zero entries. The argument above again works to force all the entries of H to be the same, unless the block C itself has a block structure

$$\left[\begin{array}{c|c} C_1 & 0 \\ \hline 0 & C_2 \end{array} \right] \tag{3.4}$$

possibly after relabeling the edge vectors and vertices. In this case, each block can have a separate m value, and the Kirchhoff graph can have disconnected subgraphs.

Finally suppose that N has an upper block of the form (3.4). Then there is a basis for Null(R) (and hence for the cycle space for our Kirchhoff graph) corresponding to the columns of N. This implies that our Kirchhoff graph is *not* vector 2-connected. □

Now consider again the tent Kirchhoff graph; it is not vector 2-connected since there are no cycles containing vectors s_1 and s_3 whose cycle vector is nonzero with respect to s_3. In terms of the current flows on this Kirchhoff graph, this lack of vector 2-connectedness implies that there is no current flow along edge vectors s_3. It is the case that the row matrix R for this example given in (3.1) is not of the form (3.4); The labels on edge vectors s_3 and s_4 must be switched to put R into the form of (3.4). So even though these two are not row equivalent, they are K-equivalent.

Remark. The discussion above proves the uniformity of vector 2-connected Kirchhoff graphs using an approach similar to that of Reese et al. [27] (2019). The main advantage to the current approach is that the digraph used here is smaller (has fewer edges) than the original one, and it more closely resembles the original Kirchhoff graph G (see example in Figure 3.2). Indeed, the desired digraph D can be obtained from G by using the same vertices (G and D have the same vertex set V), dropping the weights or edge-vector multiplicities, and reinterpreting the now single edge vectors between vertices as the directed edges of D. That the current digraph D has fewer edges than the one used in the original proof can also have computational advantages.

3.3 An implication of uniformity

One of the interesting implications of uniformity is that it allows us to look for Kirchhoff graphs with no more than a certain number of copies of each edge vector. The next proposition is a specific example of this type of result.

Proposition 3.3.1. *The Square (Figure 1.1) and Diamond (Figure 1.2) are the only multiplicity-two (m = 2) Kirchhoff graphs for any matrix row equivalent to*

$$R = \begin{bmatrix} 2 & 0 & 1 & 1 \\ 0 & 2 & 1 & -1 \end{bmatrix}$$

or the set of edge vectors in Example 1.1.

Proof. Since there can be no more than two copies of each edge vector, the only allowed vertex cuts (linear combinations of rows of R) are the following:

$$[\ 2,\quad 0,\quad 1,\quad 1\ ]$$
$$[\ 0,\quad 2,\quad 1,\ -1\ ]$$
$$[\ 1,\quad 1,\quad 1,\quad 0\ ]$$
$$[\ 1,\ -1,\quad 0,\quad 1\ ]$$
$$[\ -1,\quad 1,\quad 0,\ -1\ ]$$
$$[\ -1,\ -1,\ -1,\quad 0\ ]$$
$$[\ 0,\ -2,\ -1,\quad 1\ ]$$
$$[\ -2,\quad 0,\ -1,\ -1\ ]$$

Because a Kirchhoff graph is a finite structure (it may have no more than eight total edge vectors), the sum of the coordinates of each vertex can be taken to be nonnegative, and the origin can be taken to be as the location of the first vertex (the *anchor vertex*). Because the coordinate sums must all be nonnegative, only the first three of the eight possible vertex cuts need to be considered at the anchor vertex.

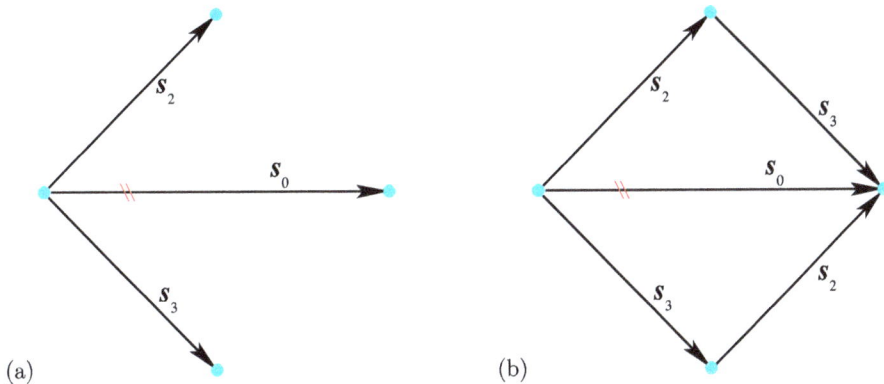

(a)   (b)

Figure 3.3: The Kirchhoff graph construction for R_1 with $m = 2$ starting with $[2, 0, 1, 1]$ at the anchor vertex (the origin). (a) Anchor vertex cut. (b) The only allowed choice for the second vertex cut at $(2, 0)$.

Suppose that the anchor vertex cut is the first on the list: $[2, 0, 1, 1]$. If this is a valid vertex cut for a Kirchhoff graph, then the Kirchhoff graph must have three additional vertices whose coordinates are $(2, 0)$, $(1, 1)$, and $(1, -1)$ (see Figure 3.3a). In addition, because $m \leq 2$, this Kirchhoff graph cannot contain any additional copies of the first edge vector s_0.

Now consider the vertex that must be at $(2, 0)$. Because there must be exactly two copies of s_0 entering this vertex, the only possible vertex cut here is $[-2, 0, -1, -1]$ (see Figure 3.3b). The other two edge vectors that this cut requires are incident on the two other vertices already required by the initial cut assigned to the anchor vertex. This cut now also uses all allowed copies of s_2 and s_3.

Finally, for the other two vertices at $(1, 1)$ and $(1, -1)$, adding the two copies of s_1 exiting from $(1, -1)$ and entering $(1, 1)$ completes a Kirchhoff graph, and indeed this is the only possible Kirchhoff graph with $m = 2$ and the first vertex cut from our list at the anchor vertex. This is the Diamond from Figure 1.2.

Now going back to the anchor vertex and assigning it the second vertex cut from our list, one finds that the process plays out essentially the same as with the first case above—it produces essentially the same Kirchhoff graph as before just translated in the plane.

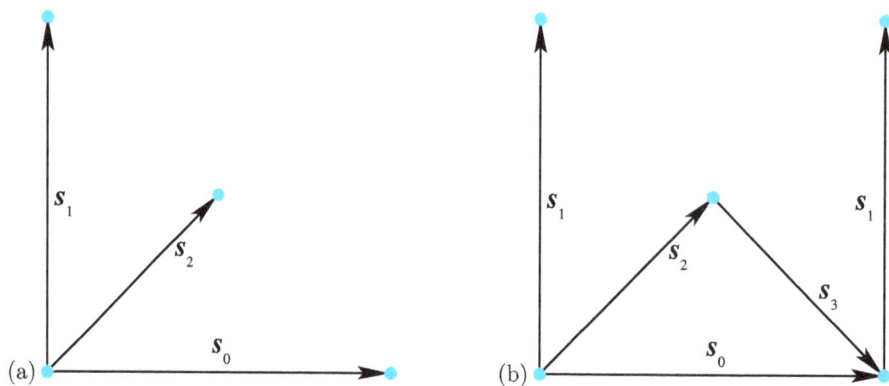

Figure 3.4: The Kirchhoff graph construction for R_1 with $m = 2$ starting with $[1, 1, 1, 0]$ at the anchor vertex (the origin). (a) Anchor vertex cut. (b) The only allowed choice for the second vertex cut at $(2, 0)$.

Finally, considering the third and final vertex cut $[1, 1, 1, 0]$ at our anchor vertex, this cut requires three additional vertices at $(2, 0)$, $(0, 2)$, and $(1, 1)$ (see Figure 3.4a). For the vertex cut at $(2, 0)$, because again $m \leq 2$, the fifth and sixth cuts from our list are the only possible choices. If the sixth cut $[-1, -1, -1, 0]$ is chosen, then all allowed copies of s_0, s_1, and s_2 have been used, and there are still four vertices to be considered. Hence if there is a Kirchhoff graph with our current anchor vertex cut, then the vertex at $(2, 0)$ must have the fifth vertex cut from our list: $[-1, 1, 0, -1]$ (see Figure 3.4b). This choice requires one new vertex at $(2, 2)$, but it also uses the last allowed copy of s_1. Still, continuing this process at the three remaining vertices shows that there is exactly one possible Kirchhoff graph that satisfies all the requirements: the Square from Figure 1.1. □

The previous proof is somewhat intricate and tedious, but this approach also can be used to find all the uniform Kirchhoff graphs for a certain set of edge vectors S or a certain row matrix R whose edge vectors appear no more times than some chosen maximum m_{max}. Although attempting this approach by hand would not be recommended, implementing it as a computer algorithm for constructing Kirchhoff graphs is possible. This algorithm is discussed in the next chapter.

4 Kirchhoff graph construction

Anyone who tries to construct a Kirchhoff graph by hand quickly realizes that the question of Kirchhoff graph construction is very much nontrivial for all but a handful of cases. For $k = 0$ (no linearly independent edge vectors) or $k = n$ (all edge vectors are linearly independent), constructing a Kirchhoff graph depends on how certain definitions are set up, but the construction is trivial. For $k = 1$ or $k = n - 1$, any Kirchhoff graph is one-dimensional or a cycle, and both these cases were dealt with in Chapter 2 (Basic results).

The interesting and often much more complicated case occurs when k is approximately $n/2$, meaning that both the row and null spaces are of similar dimension. For these cases, two algorithms have been developed that can be implemented numerically. The first, the uniformity algorithm, finds all uniform Kirchhoff graphs with edge vector multiplicity up to a certain bound. This algorithm uses the uniformity of vector 2-connected Kirchhoff graphs, as was proven in the previous chapter; it is based on the same approach as the proof of Proposition 3.3.1. The second algorithm, the linear programming algorithm, searches for a Kirchhoff graph inside a certain bounding box in the k-dimensional space using a linear programming approach. This second algorithm often finds a Kirchhoff graph more quickly, but it also finds only a few Kirchhoff graphs (perhaps only one) of some edge vector multiplicity; there is no guarantee of finding *all* Kirchhoff graphs of certain multiplicity. Given that at least one Kirchhoff graph exists for a certain set of edge vectors S, both algorithms will eventually converge to a solution, given sufficient time and computer resources.

Finally, there are a number of things to keep in mind in this chapter:

– Both algorithms presented here implicitly use Theorem 2.4.2: Each algorithm only guarantees that all the vertex cuts correspond to elements of Row(R). One must check that each vector graph produced by the algorithms actually includes all elements of S and vertex cuts that correspond to a spanning set for Row(R). If both of these conditions are met (and they typically are), then Theorem 2.4.2 guarantees that this vector graph is indeed a Kirchhoff graph.
– All Kirchhoff graphs considered in this chapter are vector 2-connected and therefore uniform.
– These Kirchhoff graphs are *prime* (formally defined below), that is, contain no Kirchhoff subgraphs.
– Both algorithms use the canonical representation of the edge vectors, implying that all the vertices in this chapter occur with integral coordinates.

https://doi.org/10.1515/9783111408576-004

4.1 Uniformity algorithm

For uniform Kirchhoff graphs, their uniformity can be a basis for an exhaustive back-tracking constructive search algorithm. This section begins with a brief overview of this algorithm, followed by a more detailed description. This algorithm is an extension of the proof method used to show that the Square and Diamond are the only Kirchhoff graphs with multiplicity two ($m = 2$) for the edge vectors in Example 1.1 (Proposition 3.3.1).

4.1.1 Uniformity algorithm overview

Recall from the Introduction that given a set of edge vectors S from some vector space V, if these vectors form only rational linear combinations, then there is a row matrix $R = [qI|C]$ whose columns are the canonical representation of the edge vectors in S and whose rows give a basis for the cut space of any Kirchhoff graph corresponding to S and R, at least for $1 < k < n$. To search for such a Kirchhoff graph, first, set up a list of all the vertex cuts that both lie in Row(R) and have entries with absolute values no greater than a given multiplicity bound m_{max}. Next, as a starting point, place an anchor vertex at the origin of the k-dimensional space, and consider whether or not the first entry on this list might be the vertex cut for this anchor vertex. If it can be, provisionally accept it and then move sequentially to each of the new vertices whose existence is required by this initial vertex cut, and ask the same question regarding this first entry from our list; if it cannot be, then move to the second entry on the list and check if it might be the vertex cut for anchor vertex. In this way, one can check all the entries from the list, first at the anchor vertex, and then at each subsequent vertex. When the final list entry is rejected at a given vertex, one goes back to the previous vertex, discards the current provisional vertex cut, and tries the next entry from the list for this vertex. The process continues until either a uniform Kirchhoff graph is found or all possible vertex cuts are rejected. Using this algorithm, for a given S, one can find all uniform Kirchhoff graphs whose multiplicities do not exceed the multiplicity bound m_{max}, or find that no such Kirchhoff graph exists. This multiplicity bound, of course, can be increased, and the search restarted.

A Java code that implements this algorithm can be found on GitHub:

https://github.com/Jessica-Wang-Math/Kirchhoff.git

This code was written by Wang [34] (2023) as a part of her Major Qualifying Project (MQP, senior thesis) under the supervision of the author, and she also helped develop the underlying algorithm. We recommend using the IntelliJ IDEA interface and inputting parameters in "Edit Configurations…".

4.1.2 Uniformity algorithm details

This section gives a more detailed description of our exhaustive backtracking constructive search algorithm (uniformity algorithm) for a given edge vector set S (or matrix R) and a given multiplicity bound m_{max}.

1. Find all possible vertex cuts by finding all linear combinations of the rows of R with entries between $-m_{max}$ and m_{max}. This will necessarily be a finite list. Let Λ be the list of these vertex cuts in an arbitrary but fixed order. Initialize $\mathbb{T}$ as an empty list for us to add potential vertices to as graph construction continues; this serves as our *to-do list*.

2. Place a starting or anchor vertex at the origin in the k-dimensional Euclidean space. Because of the way R and N are defined, all the vertices will occur at integral coordinates. In addition, because every Kirchhoff graph is finite, no vertex needs to occur at coordinates whose sum is negative. In other words, no vertex needs to be below and behind (to the left of) the anchor vertex, and our construction can begin with the southwest vertex or edge(s) of the graph.

3. Assign the first vertex cut from Λ to the anchor vertex, adding in each required edge vector and each required incident vertex. If any of these vertices has coordinates with negative sum, then delete all these edge vectors and incident vertices and consider the next vertex cut from Λ. Otherwise, add all the neighboring incident vertices to the to-do list $\mathbb{T}$.

4. Go to the next vertex v_i in the graph (according to the order in $\mathbb{T}$) and check whether it is already in Row(R). If it is not, then assign the first vertex cut from Λ to it, adding in any new edge vectors and incident vertices needed to make the vertex cut for this vertex the first entry on Λ. Check whether any of the new edge vectors and incident vertices (1) have coordinates with negative sum or (2) result in the edge vector count for any edge vector exceeding m_{max}. If either of these occurs, then delete all the new edge vectors and incident vertices and check the next vertex cut from Λ. If not, then provisionally accept this vertex cut for v_i and add all the new incident vertices to $\mathbb{T}$ if they are not already on the list. Notice that some of the new incident vertices may have previously been removed from $\mathbb{T}$, but the newly added vectors may imply that their vertex cuts are no longer in Row(R). Delete v_i from $\mathbb{T}$ and go to the next vertex on $\mathbb{T}$ after v_i.

5. When the final vertex cut on Λ is eliminated at a vertex, that vertex is abandoned, and we move back to the previous vertex, which is placed back at the top of the to-do list. For this previous vertex, we consider the next vertex cut on Λ. The new edges and incident vertices required for this next vertex cut are added, the to-do list is updated, and the process moves back to the previous step (4) in the algorithm.

6. The process ends either when there are no vertices left on the to-do list $\mathbb{T}$ (in which case a Kirchhoff graph is found, and we consider the next vertex cut from Λ at the anchor vertex to look for another Kirchhoff graph) or when the last possible vertex cut on Λ is eliminated at the anchor vertex (in which case no further Kirchhoff

graphs exist, and the entire process ends). Thus the algorithm either finds all Kirchhoff graphs G with $m(G) \leq m_{max}$, or it shows that no Kirchhoff graph exists for S with edge multiplicity not exceeding m_{max}.

7. There is one final check required to be certain that this algorithm has not found something that is not a genuine Kirchhoff graph: Does each proposed Kirchhoff graph use all the edge vectors of S (or all columns of R), and do its vertex cuts correspond to a spanning set for $\text{Row}(R)$? If the answer is "No", then the proposed Kirchhoff graph is not a genuine Kirchhoff graph. Otherwise, it is a genuine Kirchhoff graph, as it is guaranteed by Theorem 2.4.2.

4.1.3 Kirchhoff graph examples found by the uniformity algorithm

Example 4.1. As discussed in Example 1.1 and proven in Proposition 3.3.1, our uniformity algorithm shows that the two Kirchhoff graphs shown in Figure 4.1 are in fact the only ones for

$$R_1 = \begin{bmatrix} 2 & 0 & 1 & 1 \\ 0 & 2 & 1 & -1 \end{bmatrix} \tag{4.1}$$

with $m \leq m_{max} = 2$.

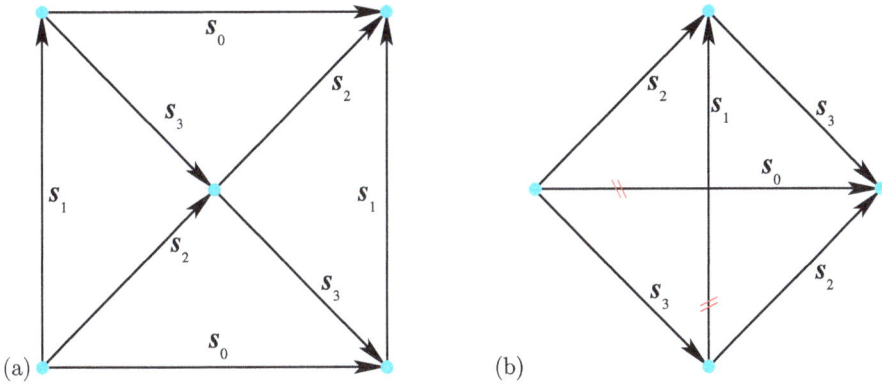

(a)   (b)

Figure 4.1: The Kirchhoff graph pair for R_1 with $m = 2$. These are the only two Kirchhoff graphs with two or fewer copies of each of the edge vectors, as was proven in Proposition 3.3.1 and confirmed using the uniformity algorithm.

Example 4.2. Given

$$R_2 = \begin{bmatrix} 2 & 0 & 1 & 1 \\ 0 & 2 & 3 & 1 \end{bmatrix} \tag{4.2}$$

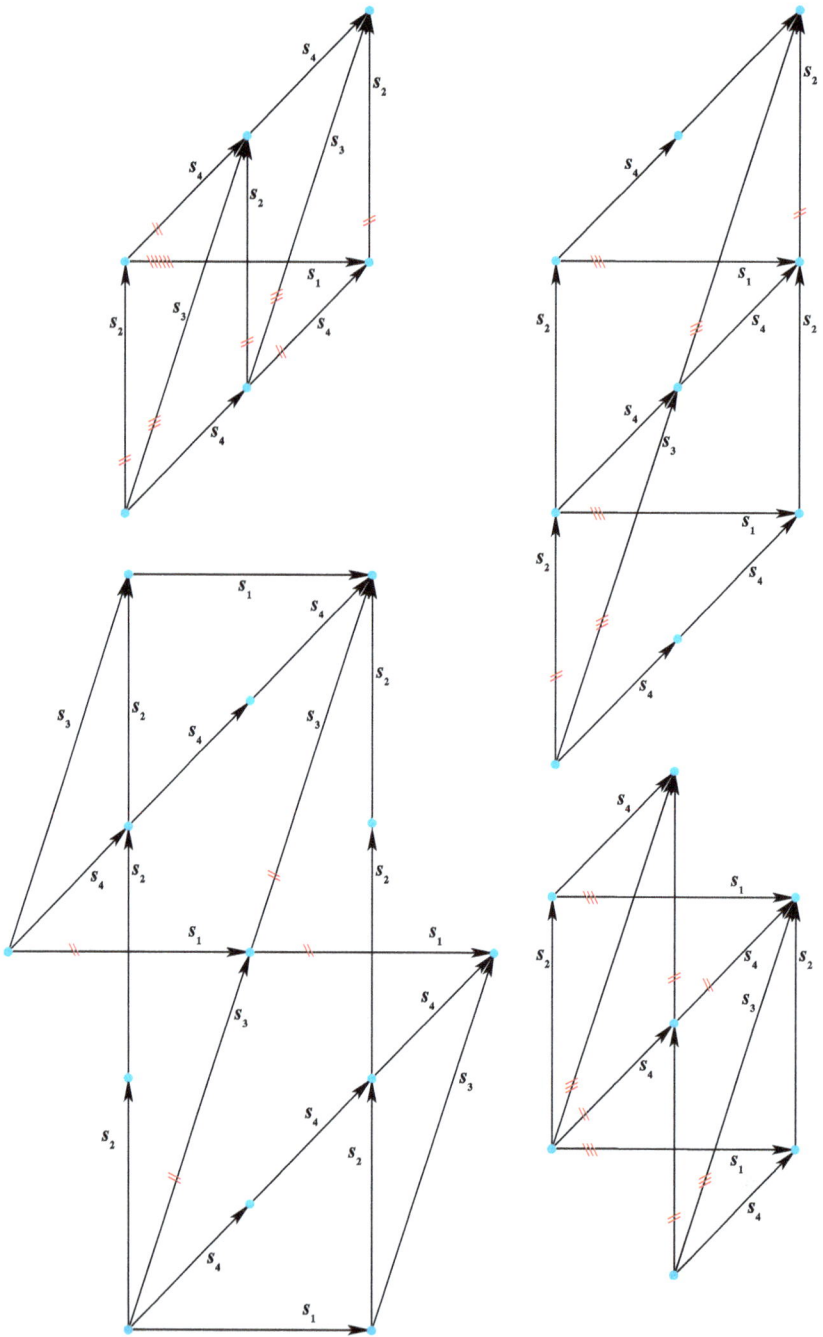

Figure 4.2: Four prime Kirchhoff graphs for the matrix R_2. These four are symmetric self-chirals. Red hash marks again give the edge multiplicity for each edge vector. Each Kirchhoff graph has six copies of each edge vector: $m = 6$ for each.

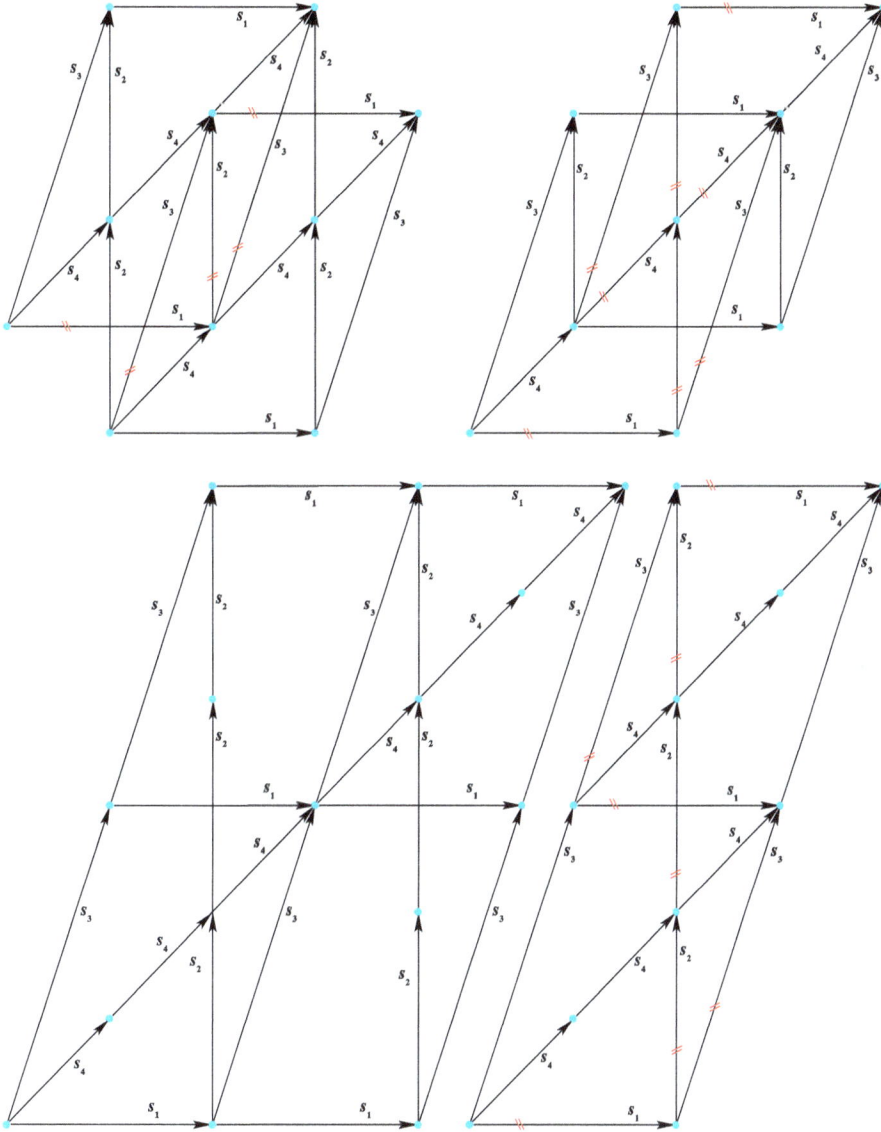

Figure 4.3: Four additional prime Kirchhoff graphs for the matrix R_2. These four are again symmetric self-chirals. Again, each Kirchhoff graph has six copies of each edge vector: $m = 6$ for each.

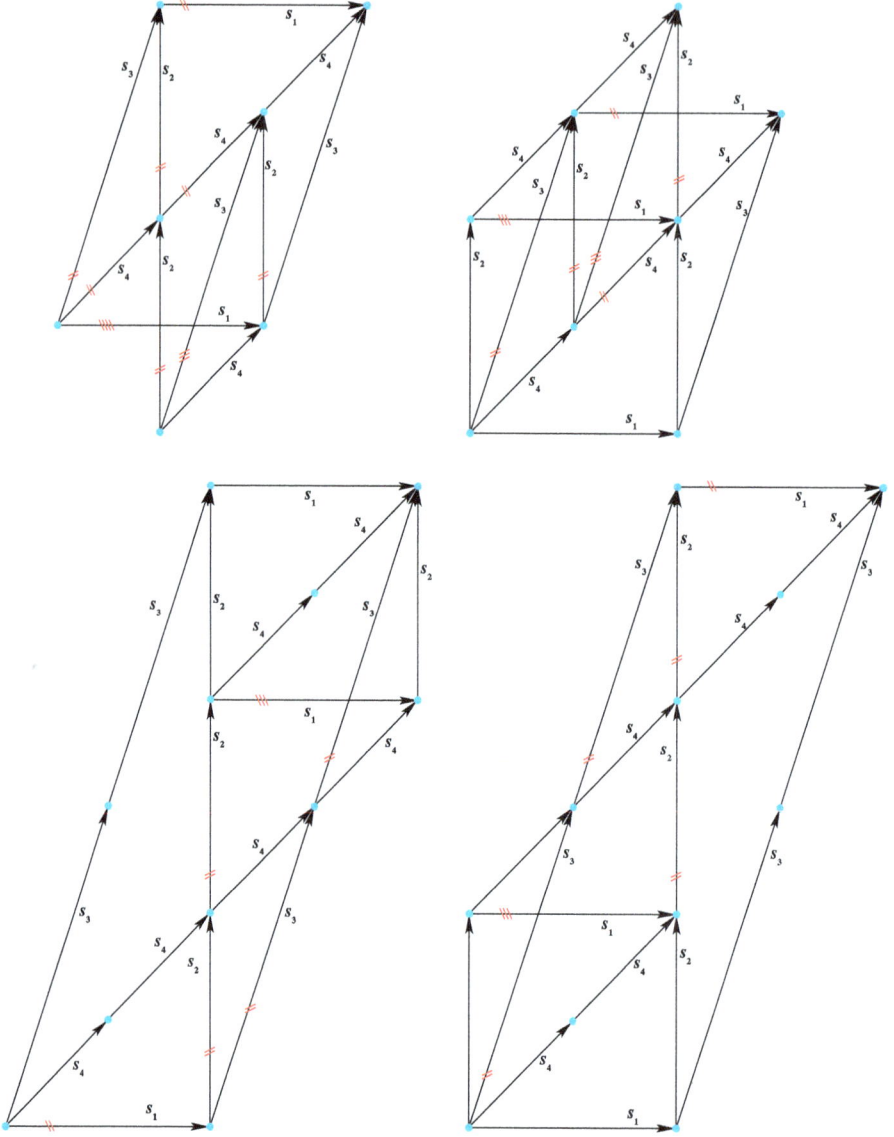

Figure 4.4: Four prime asymmetric Kirchhoff graphs for the matrix R_2, grouped in two chiral pairs. Red hash marks again give the edge vector multiplicities; there are six copies of each edge vector in each graph: again $m = 6$.

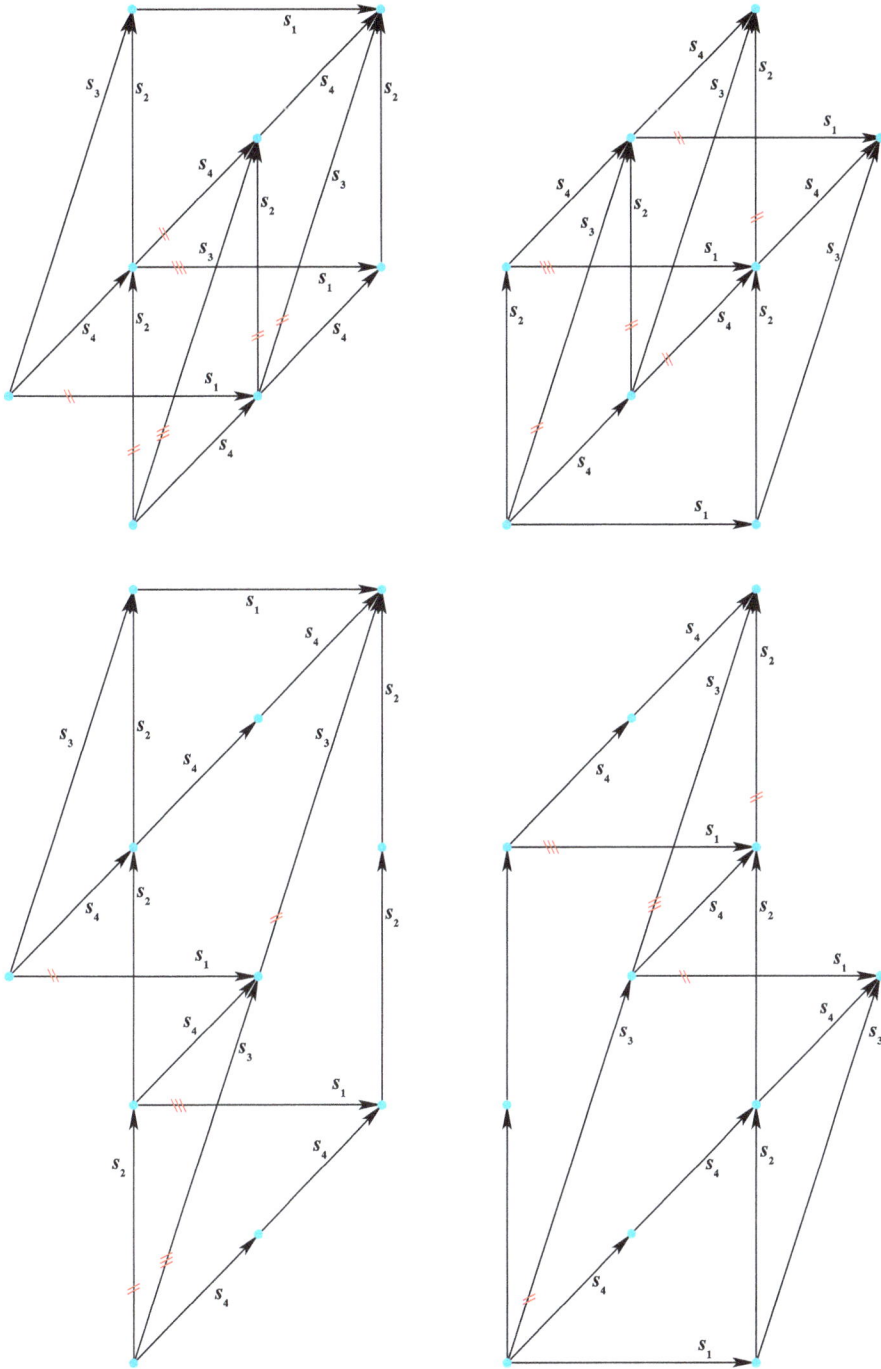

Figure 4.5: Final four prime asymmetric Kirchhoff graphs for the matrix R_2, again grouped in two chiral pairs. Again $m = 6$.

and an upper multiplicity bound $m_{max} = 6$, the algorithm finds sixteen Kirchhoff graphs, as shown in Figures 4.2–4.5. Notice that $m = 6$ is the smallest multiplicity for any Kirchhoff graph associated with R_2. The first eight Kirchhoff graphs are self-chirals, whereas the rest are chiral pairs.[1]

Example 4.3. For a third example, consider one more row matrix:

$$R_3 = \begin{bmatrix} 1 & 0 & 2 & 1 \\ 0 & 1 & 1 & 2 \end{bmatrix} \tag{4.3}$$

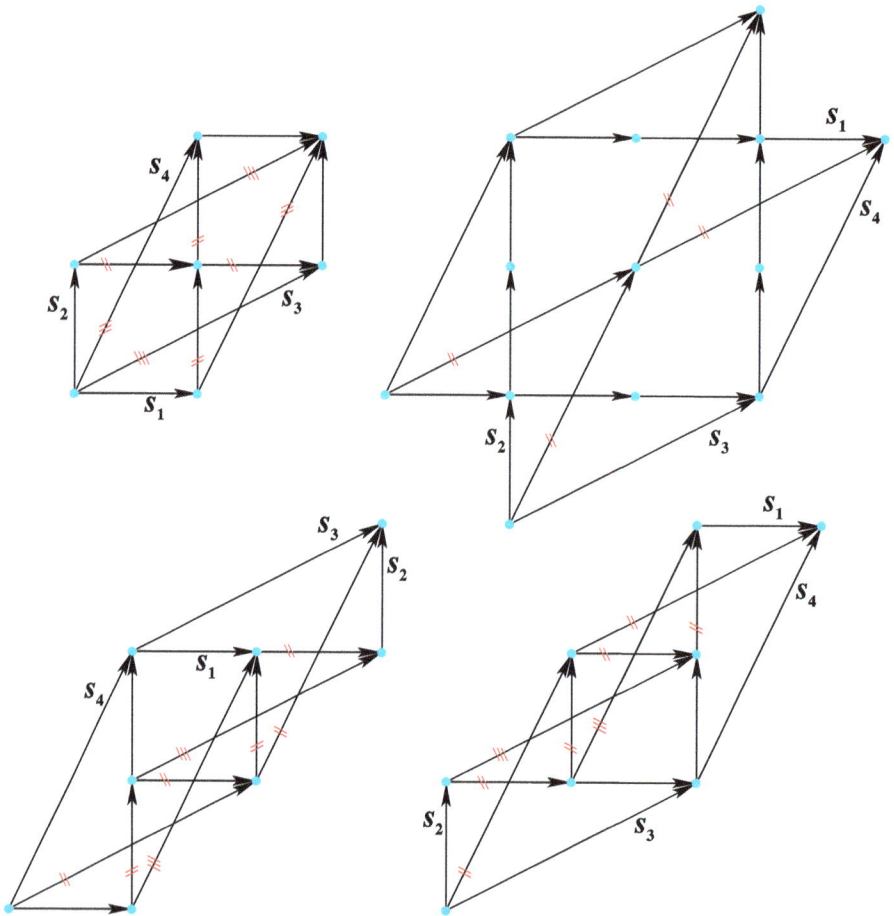

Figure 4.6: The four prime Kirchhoff graphs for the matrix R_3 having $m = 6$. Each Kirchhoff graph has six copies of each edge vector. The bottom two are a chiral pair; the top two are self-chirals.

1 As discussed in Chapter 2, two Kirchhoff graphs form a chiral pair if each is a 180-degree rotation of the other with the vectors reversed.

and again let $m_{\max} = 6$. In this case, our uniformity algorithm finds four Kirchhoff graphs with $m = 6$, as shown in Figure 4.6. Two form a chiral pair, and two are self-chirals. We can compare the two self-chiral graphs to those shown in Example 1.6 and Example 1.7. These seem to be the only four Kirchhoff graphs for R_3 with $m = 6$; indeed they must be thus if the uniformity algorithm is implemented correctly.

The implications of these three examples for the structure of families of Kirchhoff graphs will be studied more thoroughly later in this chapter, but first let us consider a second algorithm, a linear programming algorithm for finding Kirchhoff graphs.

4.2 Linear programming algorithm

Our first algorithm uses the uniformity theorem and searches for all the Kirchhoff graphs with up to a certain number of copies of the edge vectors. Of course, this algorithm requires that we have a good idea what the uniform multiplicity number is. Our second algorithm, the linear programming algorithm, does not consider the multiplicity but rather attempts to find a single Kirchhoff graph lying inside a certain bounding box in the stoichiometric space.

Suppose again that a set of edge vectors S from some vector space V is given, that these vectors form only rational linear combinations, and that there is a row matrix $R = [qI|C]$ whose columns are the canonical representation of the edge vectors in S. The mathematical issue is then again whether or not we can construct a vector graph using the vectors of S as edges, having a cycle basis corresponding to a basis for $\mathrm{Null}(R)$, and a vertex cut basis corresponding to a basis for $\mathrm{Row}(R)$. Such a vector graph G, if it exists, is a *Kirchhoff graph*.

We have already seen a number of examples of Kirchhoff graphs. Recall that in these examples, multiple copies of the same edge vector may lie on top of each other connecting the same two vertices; this multiplicity (if greater than one) has been represented by hash marks, or if larger, a weight number. Each multiplicity can also be thought of as a *weight* on that edge, and in this section, the multiplicities are referred to as weights. For each vertex, these weights are still the entries in the vertex cut vector, with positive entries indicating vectors exiting a vertex and negative entries indicating vectors entering a vertex.

The linear programming algorithm begins by establishing a bounding box and constructing the initial frame inside this bounding box. This construction is discussed in more detail below. The algorithm then determines whether or not there is an appropriate set of integer weights (copies of edge vectors) so that a Kirchhoff graph can be constructed by assigning these weights to the vectors in this initial frame. The process for determining edge weights is also described in more detail below. If a Kirchhoff graph is found inside the initial frame, the algorithm ends, and the graph is delivered; if no Kirchhoff graph is found, then the bounding box and frame are enlarged (doubled) in

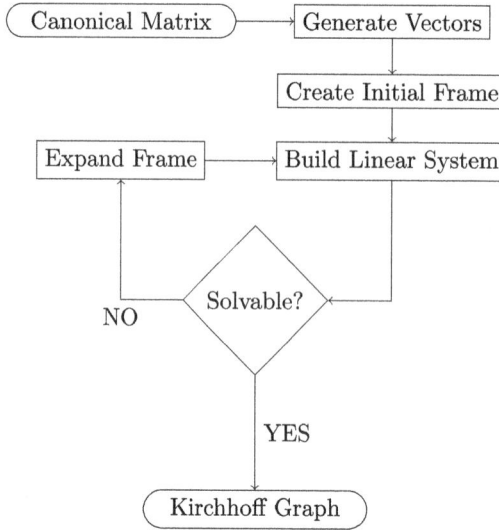

Figure 4.7: Flow chart for the linear programming algorithm. Beginning with a canonical matrix R, an initial frame is constructed inside a bounding box, then searched for a set of integer weights that satisfies (4.6) below. If such a set of weights is found, then the frame is solvable, and the Kirchhoff graph is produced. If the frame is not solvable, then the frame is enlarged in one of the coordinate directions, and a new search is made.

one of the coordinate directions, and a new search is carried out on this enlarged frame. This search process continues until either a Kirchhoff graph is found or the computer runs out of some resource (memory, storage, etc.). A flow chart for the algorithm is given in Figure 4.7.

To be specific, let us consider Example 4.4; our linear programming algorithm will be used to construct a Kirchhoff graph for this example. A second example of a Kirchhoff graph found using our linear programming algorithm is given later in this section.

Example 4.4. Consider the row matrix/null matrix pair:

$$R = \begin{bmatrix} 1 & 0 & 3 & 1 \\ 0 & 1 & 1 & 2 \end{bmatrix}, \quad N = \begin{bmatrix} 3 & 1 \\ 1 & 2 \\ -1 & 0 \\ 0 & -1 \end{bmatrix} \tag{4.4}$$

and recall that the columns of R can be used to represent the edge vectors. One Kirchhoff graph for R is shown in Figure 4.10. Again, the construction of this Kirchhoff graph using our linear programming algorithm is our current goal.

4.2.1 Initial frame construction

As discussed above, our construction process begins with a matrix R in the canonical form $R = [qI|C]$, where the n columns of R represent the set of vectors $\mathcal{S} = \{s_1, s_2, \ldots, s_n\} \subset \mathcal{V}$.

Definition. Let the k columns of qI define the **coordinate vectors** for our construction, and let the $n - k$ columns of C be the **cross-vectors**.

Our initial goal is to construct the smallest vector structure that can possibly contain a basis for Null(R) (or, equivalently, a basis for Col(N)). Consider the rectangular prism (bounding box) in $\mathbb{R}^k$ where in the ith dimension, $0 \leq x_i \leq \max_j\{R_{ij}\}$. For definiteness, let the lowest left-most corner of this prism be at the origin. The initial structure for constructing our Kirchhoff graph is a vector framework in $\mathbb{Z}^k$ sitting in this bounding box: First, place a vertex at each point in the bounding box with integral coordinates. Next, place one copy of each coordinate vector exiting each vertex $((\ell_1, \ell_2, \ldots, \ell_k) \in \mathbb{Z}^k$ where in the ith dimension, $0 \leq \ell_i < \max_j\{R_{ij}\}$). For R in (4.4), this coordinate framework is shown in Figure 4.8a. Finally, place a copy of each cross-vector exiting every vertex, *provided that the entire cross-vector remains in the prism.*

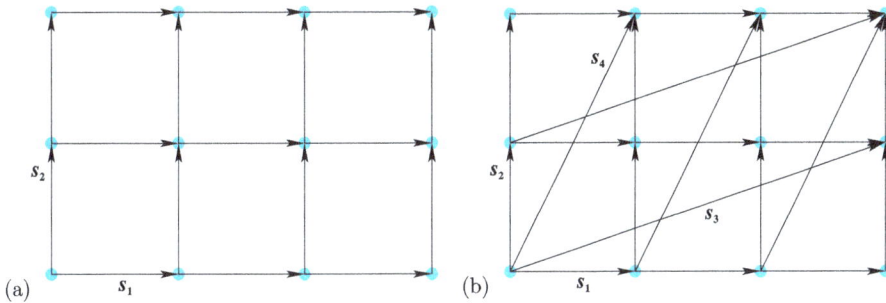

Figure 4.8: For the R, N-pair in (4.4) where $k = 2$, $\max_j\{R_{1j}\} = 3$, and $\max_j\{R_{2j}\} = 2$: (a) Coordinate framework; (b) Full initial frame for $\mathcal{S} = \{s_1, s_2, s_3, s_4\}$ based on the columns of R.

Definition. The vector structure (vector graph) described above is the **initial frame** for our Kirchhoff graph construction.

For the R, N-pair in (4.4), the initial frame is shown in Figure 4.8b. Notice that this initial frame is the smallest such structure containing a cycle basis for this R, N-pair.

4.2.2 Determining edge weights: the linear programming problem

Once a frame has been constructed (initial or enlarged), the weights for the edge vectors must be determined. Consider any vertex in the current frame (see Figure 4.9a for such a vertex in our current example). Since each vertex cut must lie in Row(R), the cut vector $[w_1, w_2, \ldots, w_n]$ for each vertex must be a linear combination of the row vectors of R:

$$[w_1, w_2, \ldots, w_n] = a_1[q, 0, 0, \ldots, 0, c_{11}, c_{12}, \ldots, c_{1,n-k}] + a_2[0, q, 0, \ldots, 0, c_{21}, c_{22}, \ldots, c_{2,n-k}]$$
$$+ \cdots + a_k[0, 0, 0, \ldots, q, c_{k1}, c_{k2}, \ldots, c_{k,n-k}]$$

for some scalars $a_i, 1 \le i \le k$. From this row space condition, $w_i = qa_i, 1 \le i \le k$. In addition, each entry in the vertex cut is the difference between the weight of a certain edge vector exiting the vertex and that of the same edge vector entering the vertex: $w_i = w_i^+ - w_i^-, 1 \le i \le n$, where w_i^+ is the weight on the exiting edge vector, and w_i^- is the weight on the entering edge vector. Combining all of the above, one finds that the $2n$ weights to be determined at this single vertex must be a solution of a linear system:

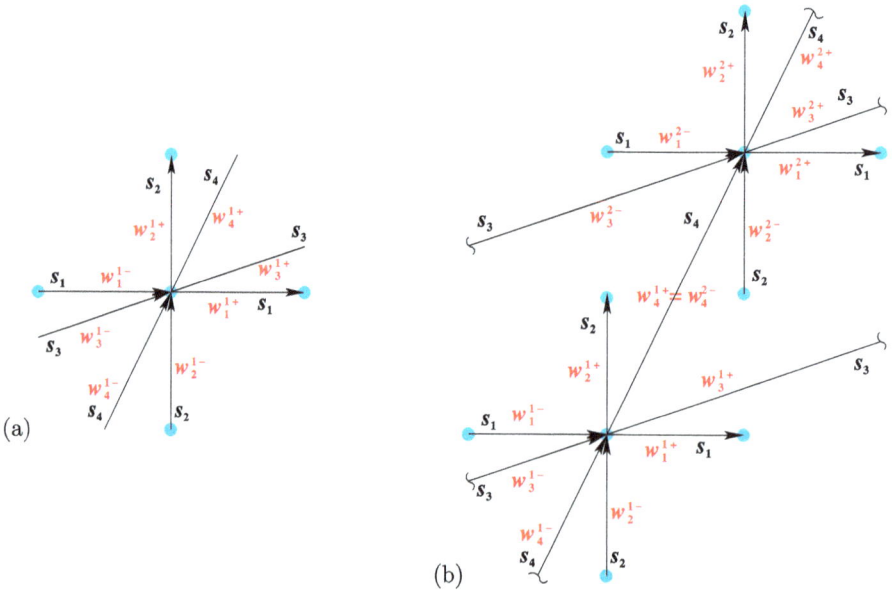

(a)

(b)

Figure 4.9: Vertices in a two-dimension frame (two coordinate vectors) with several cross-vectors. (a) The diagram on the left shows the weights for edge vectors coming into and out of a single vertex. (b) The diagram on the right shows the weights for edge vectors coming into and out of two adjacent vertices. Here $w_4^{1+} = w_4^{2-}$.

$$
\begin{bmatrix}
c_{11} & -c_{11} & \cdots & c_{k1} & -c_{k1} & -1 & 1 & 0 & 0 & \cdots & 0 & 0 \\
c_{12} & -c_{12} & \cdots & c_{k2} & -c_{k2} & 0 & 0 & -1 & 1 & \cdots & 0 & 0 \\
& \vdots & & & \vdots & & & & \vdots & & & \vdots \\
c_{1,n-k} & -c_{1,n-k} & \cdots & c_{k,n-k} & -c_{k,n-k} & 0 & 0 & 0 & 0 & \cdots & -1 & 1
\end{bmatrix}
\begin{bmatrix}
w_1^+ \\ w_1^- \\ \vdots \\ w_k^+ \\ w_k^- \\ \vdots \\ w_n^+ \\ w_n^-
\end{bmatrix}
=
\begin{bmatrix}
0 \\ 0 \\ \vdots \\ 0
\end{bmatrix}
$$

$$(4.5)$$

The same sort of linear system must now be solved for each vertex in the frame. Suppose the initial frame has M vertices; the total number of $w_i^\pm$ to be found is then $2nM$. The total number of linear equations is no more than $(3n-k)M$: there are $(n-k)M$ equations of the form given in (4.5) and up to $2nM$ additional equations because (a) the weight for any edge vector that would extend outside the frame must be zero and (b) each exiting edge weight at some vertex must match an entering edge weight at some other vertex (see Figure 4.9b). The exact number of additional equations depends on how many redundant matching conditions there are. Let $E\mathbf{w} = \mathbf{0}$ be this system (4.5). Because of the redundant matching conditions, this system is likely to have multiple nontrivial solutions, but most and maybe all of these solutions are spurious. Since all the coefficients are integers ($c_{ij} \in \mathbb{Z}$), $\mathbf{w}$ must have rational entries. Since we can multiply through by the least common multiple of the entry denominators, the weights can be given as integers. However, to correspond to a legitimate weight vector $\mathbf{w}$ for a Kirchhoff graph, all entries must also be *nonnegative*, and at least some of the entries must be positive.

To guarantee that the weights are nonnegative with some positive entries, our linear system $E\mathbf{w} = \mathbf{0}$ can be recast as a linear programming problem. Specifically, let ξ be the solution of a linear equation:

$$\mathbf{w} \cdot \mathbf{1} - \xi = 1$$

where here $\mathbf{1}$ is the vector with $2nM$ entries 1. This additional equation simply prevents the homogeneous solution $\mathbf{w} = \mathbf{0}$ from satisfying the full system. This full system is then:

$$
\begin{bmatrix} E & \mathbf{0} \\ \mathbf{1} & -1 \end{bmatrix}
\begin{bmatrix} \mathbf{w} \\ \xi \end{bmatrix}
=
\begin{bmatrix} \mathbf{0} \\ 1 \end{bmatrix}
\tag{4.6}
$$

where $w_i^\pm, \xi \geq 0$. This is a phase I linear programming problem with no objective function; such a problem may have a unique solution up to scalar multiples, multiple distinct solutions, or no solution whatsoever. If any solution is found, then the frame is deemed to be solvable, and a Kirchhoff graph can be drawn. Again, any solution will have rational weights, and these can be chosen to be integral weights by multiplying all weights by their least common denominator. Note that when any $w_i^{\pm}$ is zero, the corresponding edge vector is missing from the Kirchhoff graph.

When our linear programming algorithm is applied to the R, N-pair in Example 4.4, the initial frame is sufficient, and the Kirchhoff graph in Figure 4.10 is obtained.

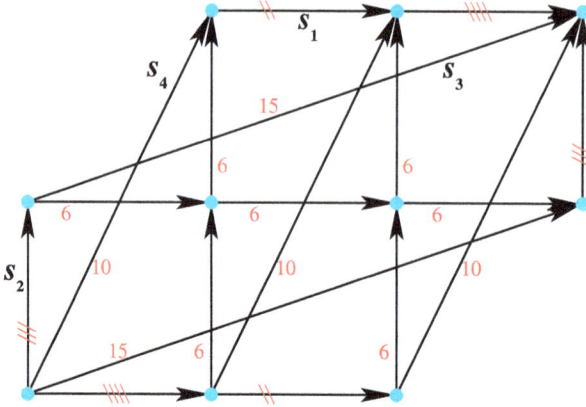

Figure 4.10: A Kirchhoff graph **G** for a set of vectors $\mathcal{S} = \{s_1, s_2, s_3, s_4\}$ whose dependencies are given by the matrices R and N in (4.4). Hash marks and Arabic numerals indicate the weights, i. e., the number of copies of a given edge vector lying in parallel and connecting the same two vertices. This depiction uses the canonical representation of the edge vectors in terms of the columns of R, and each edge vector occurs uniformly a total of thirty times: $m = 30$.

4.2.3 The linear programming code itself

The above algorithm has been implemented in Python—a readily accessible, commonly used, cross-platform coding language. The algorithm and the original version of this code was developed by Gietzmann-Sanders [12] (2017) as a part of his MQP (senior thesis) under the supervision of the author. It was later refined by Venkatraman Varatharajan as a summer undergraduate research project. The current version of this code can be found on GitHub:

https://github.com/vaakash1/kirky

The Python package in the GitHub repository provides users with a Kirchhoff object that can be used to find and investigate Kirchhoff graphs and with a drawing function to produce images. The README file and examples in the repository provide an overview of how to install and use the code, but some of these refer to an earlier version of the code. An example here may be helpful.

Example 4.5. Consider the row matrix of the form $R = [qI|C]$:

$$R = \begin{bmatrix} 1 & 0 & 0 & 2 & 1 & 1 \\ 0 & 1 & 0 & 1 & 2 & 1 \\ 0 & 0 & 1 & 1 & 1 & 2 \end{bmatrix} \tag{4.7}$$

Notice that here $q = 1$. Constructing a Kirchhoff graph by hand for this row matrix would be at best time consuming, but our linear programming code can find a Kirchhoff graph for it with little difficulty.

Once the code is downloaded from the repository, `kirky` can be installed using

```
python setup.py install
```

from a command line inside the `kirky` directory. Then `kirky` will attempt to find and draw a Kirchhoff graph corresponding to the row matrix R in (4.7) when the following lines are entered in Python:

```
from kirky import Kirchhoff
from kirky.block_q import *
from kirky.imagine import draw_graph, draw3d
import numpy as np
matrix = np.array([[2,1,1], [1,2,1], [1,1,2]])
k = Kirchhoff(matrix,q=1)
k.draw_solution(np.array([1, 0, 0.707]), np.array([0, 1, 0.707]))
```

Notice that `kirky` expects the integer values from the coefficient block C given as `matrix` and the integer value of q given in the line following the definition of `matrix`. These entries must be integers because our solutions must be computed using exact arithmetic. The values `[1, 0, 0.707]` and `[0, 1, 0.707]` determine how the three dimensions of this Kirchhoff graph are projected horizontally and vertically onto the two-dimensional page. The result will be a form of the Kirchhoff graph shown in Figure 4.11. This particular drawing is based on the output from `kirky` redrawn here to match our other Kirchhoff graphs.

Remark. The two algorithms/codes discussed here each has his own strengths and weaknesses. The great strength of the uniformity algorithm is that it finds *all* the Kirchhoff graphs of a given multiplicity. As we should see, there are important families of Kirchhoff graphs that have a common lowest possible multiplicity. The great weakness of this algorithm is that it is slow as it checks all possible vector graphs and all possible vertex cut combinations up to a given multiplicity. The search also requires significant computer resources. In contrast, the linear programming algorithm is relatively fast and can find Kirchhoff graphs with much larger multiplicities. The weakness of this

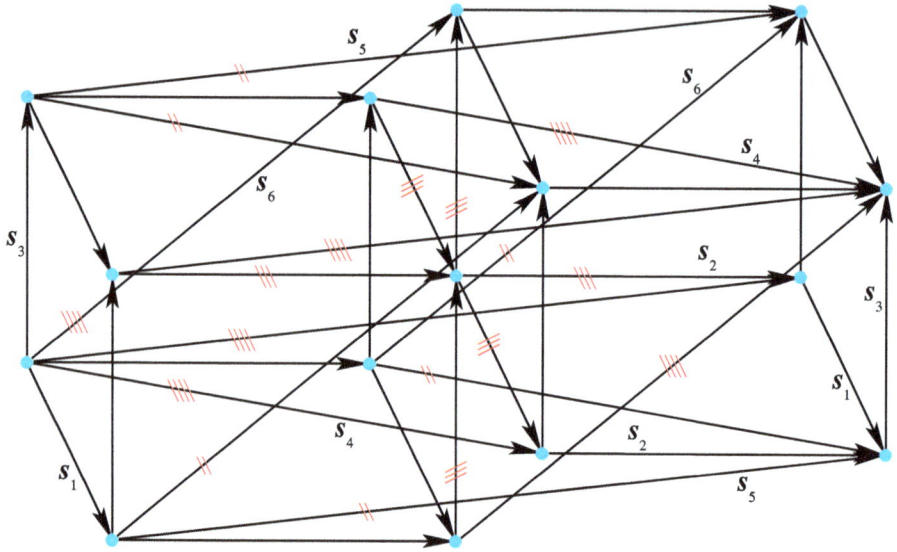

Figure 4.11: A Kirchhoff graph for the row matrix R in (4.7) and the set of vectors $S = \{s_1, s_2, s_3, s_4, s_5, s_6\}$ based on the columns of R. Hash marks indicate the weights, i. e., the number of copies of a given edge vector lying in parallel and connecting the same two vertices. Each edge vector occurs uniformly a total of twelve times.

code is that it finds a single Kirchhoff graph or perhaps several Kirchhoff graphs, but it cannot be expected to find all the Kirchhoff graphs of some meaningful size.

The issues discussed in the previous remark are illustrated by the following example. As it turns out, the smallest m value for which there is a Kirchhoff graph is $m = 30$. This multiplicity is too large for our uniformity algorithm to handle. The linear programming algorithm, on the other hand, can find some Kirchhoff graphs but likely only a portion of the total.

Example 4.6. Consider another row-null matrix pair:

$$R = \begin{bmatrix} 1 & 0 & -1 & 2 \\ 0 & 1 & 1 & 3 \end{bmatrix}, \quad N = \begin{bmatrix} -1 & 2 \\ 1 & 3 \\ -1 & 0 \\ 0 & -1 \end{bmatrix} \tag{4.8}$$

A symmetric prime Kirchhoff graph for the pair in (4.8) is shown in Figure 4.12. This Kirchhoff graph was originally found by hand, but also found by our linear programming algorithm. Two asymmetric prime Kirchhoff graphs for this pair (4.8) are shown in Figure 4.13. These two asymmetric Kirchhoff graphs are a chiral pair that would be difficult or impossible to find by hand. Both can be seen to be prime since the edge vector s_4 ($[2, 3]^T$) must be in each graph to ensure that it has a cycle basis and since the

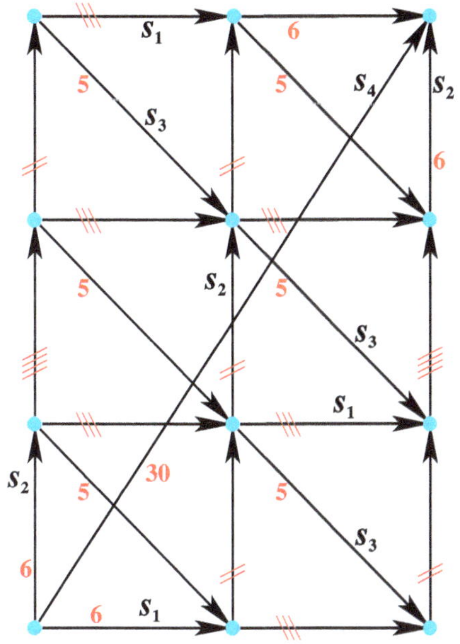

Figure 4.12: A symmetric prime Kirchhoff graph for R and N in (4.8). Here $m = 30$; indeed, there are thirty copies of the edge vector s_4 between the lower left and upper right, making this a Kirchhoff strut graph.

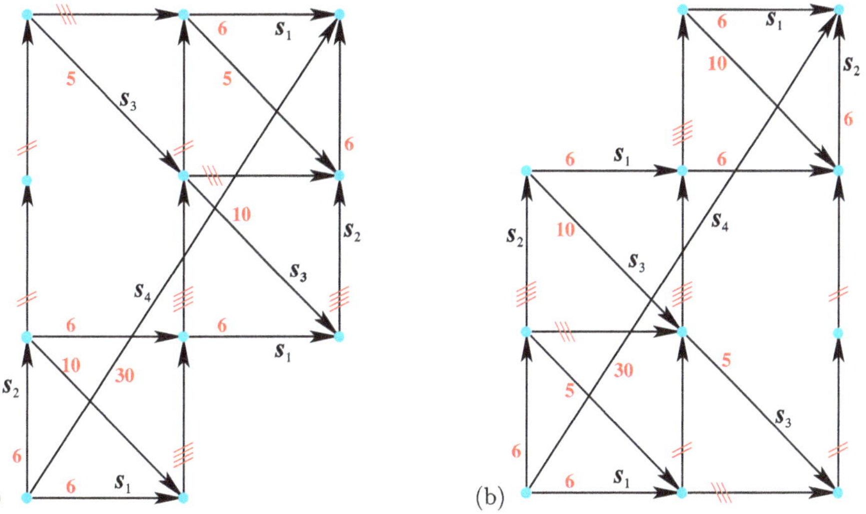

Figure 4.13: (a) An asymmetric Kirchhoff graph for R and N in (4.8) found by our linear programming algorithm. (b) The chiral of (a). Both are prime. Finding asymmetric Kirchhoff graphs by hand would be extremely difficult. Each Kirchhoff graph again has thirty copies of each edge vector, and again this is a Kirchhoff strut graph.

uniform number of each edge vector must be divisible by 2, 3, and 5 because all vertices not involving s_4 must have a vertex cut that is a multiple of $[3, -2, -5, 0]$. This means that there must be at least thirty copies of each edge vector in any Kirchhoff graph for this matrix pair. Since all thirty copies of s_4 form a sort of strut disgonally across these Kirchhoff graphs, these are considered a strut Kirchhoff graph

4.3 Results and conjectures found using the algorithms

As mentioned above, our algorithms allow us to construct larger and more complicated Kirchhoff graphs than are achievable by hand, but there are a number of other results and conjectures that can be obtained at least in part using these algorithms. First, there are a number of definitions and basic results that need to be introduced or reviewed here.

Definition. A Kirchhoff graph is **trivial** if it contains one vertex and no edge vectors. A nontrivial Kirchhoff graph G is **prime** if and only if no nontrivial Kirchhoff subgraph can be obtained by removing one or more edge vectors from G. A nontrivial Kirchhoff graph is **composite** if it is not prime. A **chiral** of a Kirchhoff graph is obtained by taking a two-dimensional projection of a Kirchhoff graph (if the graph was not initially embedded in the plane), rotating it by 180°, and then reversing each individual edge vector. A Kirchhoff graph is **symmetric** if it is self-chiral, that is, its chiral is itself. A Kirchhoff graph is **asymmetric** if it is not symmetric. (If either $R = [0]$ or $R = I$, then take the Kirchhoff graph to be trivial, that is, a single vertex with no edge vectors.)

An immediate result is just a restatement of Proposition 2.3.1:

Proposition 4.3.1. *The chiral of any Kirchhoff graph is itself a Kirchhoff graph.*

The next two propositions require the concept of *tiling* that in turn requires the operations of addition and subtraction of vector graphs (often Kirchhoff graphs).

Definition. Given two vector graphs G_1, G_2, both based on a edge vector set S, and a coordinate $x \in \mathbb{Z}^k$, the **sum** $(G_1 + G_2, x)$ is defined as the union of G_1 with its anchor vertex placed at the origin and G_2 with its anchor vertex placed at the coordinate x. Then for the vertices, $V(G_1 + G_2) = V(G_1) \cup V(G_2)$, and for the multiplicities, $m(G_1 + G_2) = m(G_1) + m(G_2)$. Thus *all* copies of the edge vectors from both G_1 and G_2 are present in their sum. When there is no ambiguity, write $G_1 + G_2$ as a shorthand for $(G_1 + G_2, x)$. In addition, define the **difference** $(G_1 - G_2, x)$ or $G_1 - G_2$ as the removal of G_2 from G_1 assuming that there is a complete copy of G_2 as a subgraph of G_1 anchored at x. Collectively, the process of repeatedly adding and/or subtracting Kirchhoff graphs is called **tiling**, and the resulting Kirchhoff graph is a **tiling** of the original Kirchhoff graphs.

Proposition 4.3.2. *Provided that the sum or difference is cyclic, the sum or difference of two Kirchhoff graphs G_1 and G_2 is itself a Kirchhoff graph.*

Proof. Since G_1 and G_2 both have vertex cuts that span the cut space and cycles that span the cycle space, so must $G_1 + G_2$. (Theorem 2.4.2 guarantees that the sum cannot have any additional cycles.) For $G_1 - G_2$, since edge vectors are removed from G_1, the difference $G_1 - G_2$ must have the same cycle space as G_1 or a smaller one. The same is true for the cut space: since G_1 and G_2 have the same cut space, the cut space of $G_1 - G_2$ must be the same or smaller, not larger. Then again by Theorem 2.4.2, neither space can get smaller without the other getting larger. Thus $G_1 - G_2$ has the same cycle space and cut space as the two original Kirchhoff graphs, and hence it is also a Kirchhoff graph. □

Proposition 4.3.3. *Any composite Kirchhoff graph C is obtained by tiling two or more prime Kirchhoff graphs.*

Proof. Since C is a composite Kirchhoff graph, $C = G_1 + V$, where G_1 is Kirchhoff graph, and V is a vector graph. Then $V = C - G_1$ must itself be a Kirchhoff graph, and V may be either prime or composite. If it is composite, then restart this argument to find G_2. Since these graphs are finite structures, this process must eventually terminate with a set of prime Kirchhoff graphs. □

A couple examples of Kirchhoff graph addition/tiling should be helpful:

Example 4.7. A simple example of prime and composite Kirchhoff graphs can be seen in Figure 4.14. These Kirchhoff graphs are associated with a specific R,N-pair of matrices:

$$R = \begin{bmatrix} 1 & 0 & 1 \\ 0 & 1 & 1 \end{bmatrix}, \quad N = \begin{bmatrix} 1 \\ 1 \\ -1 \end{bmatrix} \tag{4.9}$$

The two triangular Kirchhoff graphs in Figure 4.14a are the two smallest prime chirals for (4.9); they are of course asymmetric. One simple composite tiling of these two triangular Kirchhoff graphs is shown in Figure 4.14b.

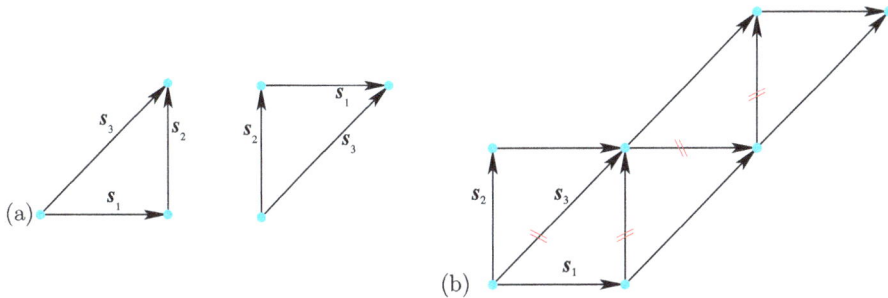

Figure 4.14: (a) Two prime Kirchhoff graphs that are chirals of each other. (b) A composite Kirchhoff graph, a tiling of the two prime triangles.

Example 4.8. Consider again Example 1.1. Let the Square Kirchhoff graph in Figure 4.1a be denoted F_1, and let the Diamond Kirchhoff graph in Figure 4.1b be denoted F_2. Both these Kirchhoff graphs are prime because there are no smaller nontrivial Kirchhoff graphs for this edge vector set S. Then their sum $(F_1 + F_2, (1, 1))$ is shown in Figure 4.15. This sum is definitely a composite Kirchhoff graph.

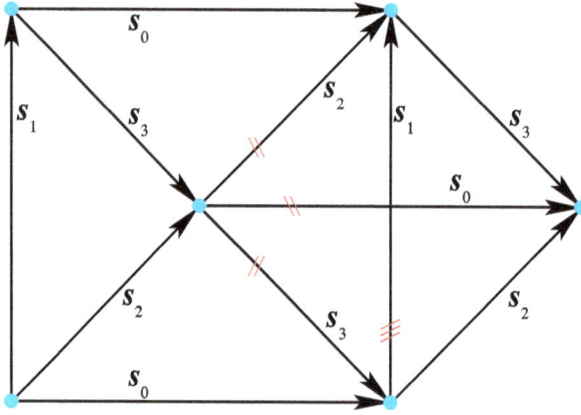

Figure 4.15: An addition example $(F_1 + F_2, (1, 1))$ for the row matrix R_1 in (4.1).

Tiling Kirchhoff graphs in various ways can generate a wide variety of larger Kirchhoff graphs, far wider than the two simple examples above suggest. This variety can be seen in the next subsection.

4.3.1 Fundamental graphs and Kirchhoff graph families

This subsection considers the structure of Kirchhoff graph families based on their edge vector multiplicity. The key set of Kirchhoff graphs to consider is the family of all Kirchhoff graphs generated by tiling:

Definition. Given N Kirchhoff graphs $G_1, \ldots, G_N$ each based on the same edge vector set S, the **family of Kirchhoff graphs generated by tiling** copies of these graphs is $\langle G_1, \ldots, G_N \rangle := \{a_1 G_1 + \cdots + a_N G_N : a_i \in \mathbb{Z}\}$, where the sum/difference is taken over all choices of a_i and all anchor vertex placements, provided that the resulting graphs remain cyclic. The a_i can be negative only when there exists $|a_i|$ copies of G_i in the larger Kirchhoff graph.

Proposition 4.3.4. *A Kirchhoff graph G is prime if and only if G has no nontrivial Kirchhoff subgraph decomposition. In other words, G cannot be written as $G_1 + G_2$, where both G_1 and G_2 are nontrivial Kirchhoff graphs.*

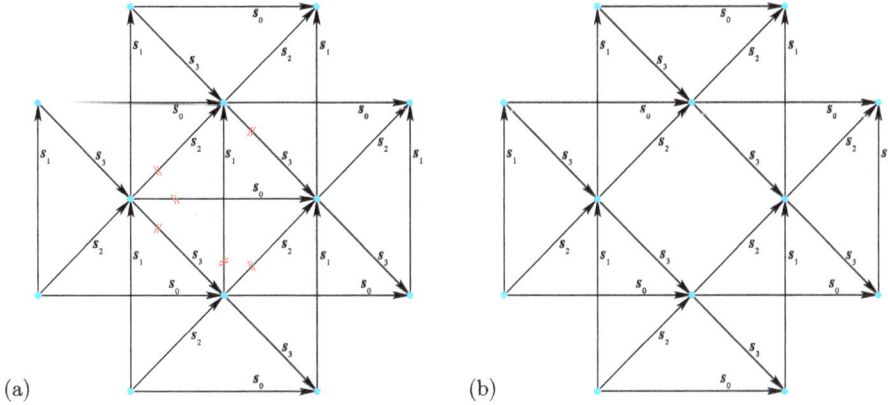

Figure 4.16: The Kirchhoff graph on the left (a) is the composite $C_1 := 4F_1$; the one on the right (b) is the prime $P_1 := 4F_1 - F_2$.

The Kirchhoff graphs shown in Figures 4.1–4.6, 4.10, and 4.11 are all prime; the one in Figure 4.15 is of course composite. It might seem that all the Kirchhoff graphs in $\langle F_1, F_2 \rangle$ other than F_1 and F_2 themselves must be composite. Perhaps surprisingly, this is not the case: Consider the two Kirchhoff graphs in Figure 4.16. The Kirchhoff graph on the left is $C_1 := 4F_1$ (a tiling by addition of four copies of $F1$); it is clearly composite. The Kirchhoff graph on the right is $P_1 := 4F_1 - F_2$ (the copy of F_2 in the middle has been removed); it is prime. This follows since if any edge vector is removed from P_1, the vertex cuts for incident vertices will no longer lie in $\text{Row}(R_1)$, and the only way to return all the vertex cuts to $\text{Row}(R_1)$ by removing additional edge vectors is to remove all the edge vectors, meaning that there is no Kirchhoff subgraph.

The above example makes clear that composite Kirchhoff graphs may not have unique prime decompositions. In this example, $C_1 = 4F_1 = P_1 + F_2$. On the other hand, there are infinitely many prime Kirchhoff graphs in $\langle F_1, F_2 \rangle$:

Proposition 4.3.5. *Let F_1 and F_2 be the two prime Kirchhoff graphs for the matrix*

$$R_1 = \begin{bmatrix} 2 & 0 & 1 & 1 \\ 0 & 2 & 1 & -1 \end{bmatrix}$$

with $m = 2$ (see Figure 4.1). The family $\langle F_1, F_2 \rangle$ contains infinitely many prime Kirchhoff graphs.

Proof. Constructing arbitrarily large prime Kirchhoff graphs that are in $\langle F_1, F_2 \rangle$ is a simple extension of the construction of the prime Kirchhoff graph in Figure 4.16b. Simply form $6F_1 - 2F_2$ by adding two more copies of F_1 to the prime Kirchhoff graph P_1 in Figure 4.16 and then by subtracting F_2 from the middle. Notice that the addition is exactly what is needed to create the copy of F_2 and thus make possible the subtraction. We can

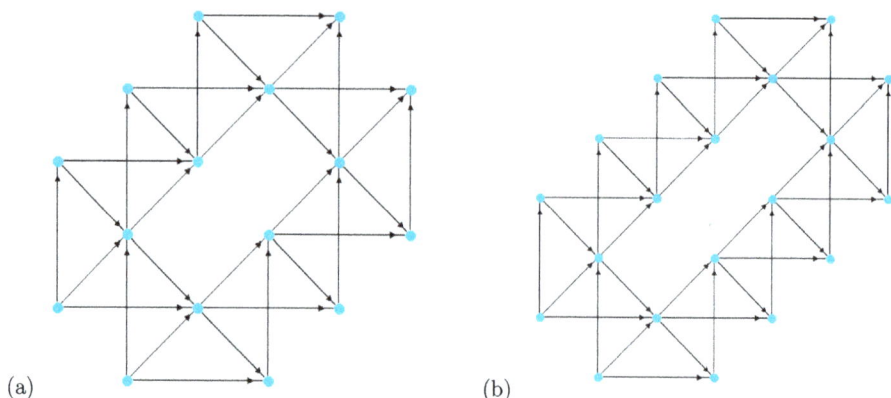

(a) (b)

Figure 4.17: Two larger prime Kirchhoff graphs generated by tiling for R_1. On the left, $P_2 = 6F_1 - 2F_2$; on the right, $P_3 = 8F_1 - 3F_2$.

repeat this tiling until the prime Kirchhoff graph of the desired size is achieved. The next two such prime Kirchhoff graphs in $\langle F_1, F_2 \rangle$ are shown in Figure 4.17. □

Remark. Although it is hard to describe in general, this sort of construction of prime Kirchhoff graphs of arbitrary size seems to be possible in $\langle K_1, K_2 \rangle$ for any pair of prime Kirchhoff graphs of minimal multiplicity for a given row matrix R. This is particularly true when the generating Kirchhoff graphs K_1 and K_2 are symmetric (self-chiral).

Another important outcome of these algorithms is a beginning of an understanding of the structure of the families generated by tiling Kirchhoff graphs. Notice that for a given set of edge vectors S and the corresponding row matrix R, assuming that a Kirchhoff graph exists, there is a minimum edge multiplicity number m^* below which there can be no nontrivial Kirchhoff graphs. For $m = m^*$, there will be a finite number of Kirchhoff graphs, and each of these will be prime since any proper Kirchhoff subgraph would have fewer than m^* edge vectors. It is frequently the case, however, that one or more of these graphs are in fact tilings of the others. An example of this is shown in Figure 4.6 for R_3, where any one of the four Kirchhoff graphs is a tiling of the other three. Specifically, if we add the two symmetric Kirchhoff graphs at the top of Figure 4.6 so that the null vertices at the center of each graph are identified and then subtract the asymmetric graph at the lower left in Figure 4.6, then the result is the asymmetric graph at the lower right. This structure leads to an additional definition, that of *fundamental Kirchhoff graphs*.

Definition. A **fundamental set** for R and S is a minimal generating set in terms of tiling with respect to both multiplicity and cardinality. Members of a fundamental set are called **fundamental Kirchhoff graphs**.

For R_1 in (4.1), both Kirchhoff graphs in Figure 4.1 are fundamental. For R_3 in (4.3), any three of the Kirchhoff graphs in Figure 4.6 are fundamental. For R_2 in (4.2), no more than twelve of the sixteen are fundamental. Based on current computations, all known larger Kirchhoff graphs are tilings of the graphs in the fundamental set, that is, all are in $\langle F_1, F_2, \ldots, F_N \rangle$, and thus all have an edge vector multiplicity that is an integral multiple of m^*, though proof of this result remains open.

4.3.2 Conjecture on the value of m^*

One of the interesting questions that our computations raise is whether or not we can compute the value of m^* based on either the edge vectors $\mathcal{S}$ or the row matrix R without actually constructing the family of Kirchhoff graphs. Although proving that the following method actually produces m^* is still an open question, the method does seem effective in determining its value. The method was found based on numerical experiments.

Conjecture 4.3.1. *For a given row matrix of the form $R = [qI|C]$, recall that I is the $k \times k$ identity matrix, so each row of R has at least $k - 1$ zero entries. Notice that through row operations, we can find equivalent rows to those in R where the $k - 1$ zero entries occur in any of the n columns of R. Let the matrix R^* be row-equivalent to R with R as the top block and equivalent rows with all possible distributions of the $k - 1$ zeros among the n columns below.[2] The value of m^* is then conjectured to be the least common multiple of the absolute values of the entries of R^*:*

$$m^* = \mathrm{lcm}\left(|R^*_{ij}|\right)$$

Example 4.9. Consider the following row matrix

$$R = \begin{bmatrix} 1 & 0 & 2 & 7 \\ 0 & 1 & 3 & 1 \end{bmatrix} \tag{4.10}$$

that is row-equivalent to an extended matrix:

$$R^* = \begin{bmatrix} 1 & 0 & 2 & 7 \\ 0 & 1 & 3 & 1 \\ -3 & 2 & 0 & -19 \\ 1 & -7 & -19 & 0 \end{bmatrix} \tag{4.11}$$

The least common multiple then suggests that $m^* = 2 \cdot 3 \cdot 7 \cdot 19 = 798$. Using the linear programming algorithm, one finds a prime Kirchhoff graph indicating that this m^* value

---

2 Depending on the entries in C, there may be more than $k - 1$ zeros in some rows. If extra zeros occur in rows, then this simply reduces the total number of rows in R^*.

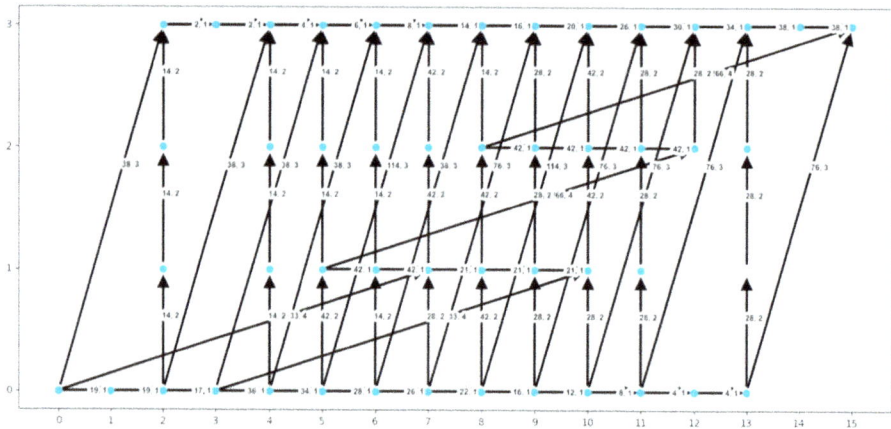

Figure 4.18: A prime Kirchhoff graph for R in (4.10). Here $m^* = 798$, so this is the smallest possible multiplicity for a Kirchhoff graph with edge vectors given by the columns of this R. Each edge vector is labeled with a label of the form x, y, where x is the edge vector multiplicity or weight, and y is the edge vector index.

is correct in this case (see Figure 4.18). This Kirchhoff graph can be shown to be prime by removing any sequence of edge vectors in an attempt to find a Kirchhoff subgraph. In the end, the only possible Kirchhoff subgraph is the trivial one. There are at least 29 distinct prime Kirchhoff graphs for this R; all found so far are asymmetric.

The previous example raises another question: Given a row matrix R and minimal multiplicity m^*, how many prime Kirchhoff graphs are there with $m = m^*$, and how to determine this value? While no general answer exists for this question, it is worth noting that the value can vary significantly from row matrix to row matrix. From some of the previous examples in this chapter we found that for R_1 (4.1), there are two prime Kirchhoff graphs with $m = m^* = 2$, both of which are fundamental. For R_2 (4.2), there are sixteen prime Kirchhoff graphs with $m = m^* = 6$, and no more than twelve are fundamental. For R_3 (4.3), there are four prime Kirchhoff graphs again with $m = m^* = 6$, any three of which are fundamental. The following example shows that the number of prime Kirchhoff graphs with $m = m^*$ can vary even more widely than the previous three examples suggest.

Example 4.10. Consider the row matrix

$$R = \begin{bmatrix} 1 & 0 & 1 & 3 \\ 0 & 1 & 1 & -1 \end{bmatrix}. \tag{4.12}$$

For this row matrix R, $m^* = 12$, and for $m = m^*$, there are 131 prime Kirchhoff graphs, 19 being symmetric, whereas 112 are asymmetric appearing in 56 chiral pairs. The most compact of these prime Kirchhoff graphs is shown in Figure 4.19.

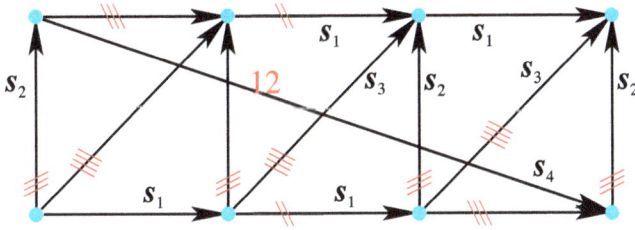

Figure 4.19: A most compact prime Kirchhoff graph for R in (4.12). Here $m = 12$; there are 130 other prime Kirchhoff graphs for this R, all with $m = 12$. Again this is a strut Kirchhoff graph.

It is difficult to say why there should be two smallest prime Kirchhoff graphs for one row matrix, sixteen for another, and then 131 for a third. This also remains an open question.

Another open question is whether or not there are any Kirchhoff graphs for a given set of edge vectors S that are *not* in $\langle F_1, F_2, \ldots, F_N \rangle$, the family of Kirchhoff graphs generated by a fundamental set of prime Kirchhoff graphs whose multiplicity is m^*. So, for example, the Square (F_1) and the Diamond (F_2) in Example 4.1 form a fundamental set for R_1 in (4.1). In this case, $m^* = 2$, meaning that all of the Kirchhoff graphs in $\langle F_1, F_2 \rangle$ have even multiplicities. Could there be a Kirchhoff graph for this S with, say, $m = 73$? Numerical computations suggest that this is not the case, that all the Kirchhoff graphs for S are in the family of Kirchhoff graphs generated by the Square and the Diamond, but of course, these computations are no proof.

Remark. It is perhaps true that calling the process of building larger Kirchhoff graphs from smaller ones "tiling" is not ideal. The author likes to think of the process as akin to snapping Legos together. The interesting slant on this view is that we can start with four identical Legos, snap them together in just the right way, then pop out a Lego from the assemblage different from the four we started with, and be left with yet another distinct Lego piece. "Tiling" has the advantage of simplicity.

5 Matroids and Kirchhoff graphs over finite fields

Kirchhoff graphs and matroids share at their foundations one central connection: they are both concerned with the dependencies among the vectors from some set of vectors. This chapter tries to make this connection clear. It starts with a short introduction to matroids, then discusses the Cayley color graph, and shows how it can be used to construct $\mathbb{Z}_p$-Kirchhoff graphs for vectors or matrices over finite fields and for binary matroids. This explicit construction of Kirchhoff graphs over finite fields is in stark contrast to the situation for Kirchhoff graphs over the rationals where the existence is still an open question. In particular, Theorem 5.3.1 shows that for any integer-valued matrix A, there exists a nontrivial $\mathbb{Z}_p$-Kirchhoff graph for A (mod p) for sufficiently large prime p. Moreover, the proof of Theorem 5.3.2 gives an explicit construction of a nontrivial $\mathbb{Z}_p$-Kirchhoff graph for $\mathbb{Z}_p$-valued matrices with an entrywise nonzero vector in the row space. Finally, Section 5.4 demonstrates that every binary matrix or binary matroid has a nontrivial Kirchhoff graph, that is, a matrix representation of any binary matroid has an appropriate Kirchhoff graph and is thus *Kirchhoff graphic*. In some sense, this result follows since the same edge vector can appear in multiple places in a Kirchhoff graph, even though edges in a (standard) graph are always distinct.

5.1 Matroids

The concept of a matroid was originally introduced by Whitney [37] (1933). Our discussion here focuses only on what is needed to make connections with Kirchhoff graphs. For a general introduction to matroids, see, for example, Welsh [36] (1976), Oxley [19] (2011), or Pitsoulis [21] (2014).

There are a number of equivalent definitions for matroids; the one involving independent sets is most useful for our purposes:

Definition. Let S be a finite set of elements (the **ground set**), and let $\mathcal{I}$ be a set of independent subsets of S (the **family of independent sets**). The **matroid** $M(S, \mathcal{I})$ is the pair $(S, \mathcal{I})$ satisfying the following three conditions:

1. **Empty set condition**: The empty set is independent: $\emptyset \in \mathcal{I}$
2. **Subset condition**: Every subset of an independent set is independent: If $B \subset A \subset S$ and $A \in \mathcal{I}$, then $B \in \mathcal{I}$.
3. **Augmentation condition**: If A and B are two independent sets and A has more elements than B, then there exists an element of A that is not in B such that the set containing exactly this element and the elements of B is independent: If $A, B \in \mathcal{I}$ and $|A| > |B|$, then there is $x \in A - B$ such that $\{x\} \cup B \in \mathcal{I}$.

Any subset of S that is not *independent* is called a **dependent set**. A minimal dependent set is a **circuit**.

https://doi.org/10.1515/9783111408576-005

Example 5.1. Returning to Example 1.1 in the Introduction, let the ground set be $S = \{s_0, s_1, s_2, s_3\}$, where again

$$s_0 = \begin{bmatrix} 2 \\ 0 \end{bmatrix}, \quad s_1 = \begin{bmatrix} 0 \\ 2 \end{bmatrix}, \quad s_2 = \begin{bmatrix} 1 \\ 1 \end{bmatrix}, \quad s_3 = \begin{bmatrix} 1 \\ -1 \end{bmatrix}$$

and there is a natural family of independent sets:

$$\mathcal{I} = \{\emptyset\} \cup \{\text{all singleton subsets of } S\} \cup \{\text{all paired subsets of } S\}$$

Since any set of three of these vectors has a linear combination that adds to zero and is thus represented by a cycle in the two Kirchhoff graphs in Example 1.1, no larger subsets can be included in $\mathcal{I}$. Therefore it is relatively easy to check that this ground set S and family of independent sets $\mathcal{I}$ satisfy the three conditions, making $(S, \mathcal{I})$ a matroid.

Example 5.2. The previous example is perhaps a bit too simple because no three vectors can be independent in $\mathbb{R}^2$, the natural embedding for that Kirchhoff graph. As a somewhat more interesting case, consider the following example: Let the ground set be $S = \{s_1, s_2, s_3, s_4, s_5\}$ with the following s_i:

$$s_1 = \begin{bmatrix} 1 \\ 0 \\ 0 \end{bmatrix}, \quad s_2 = \begin{bmatrix} 0 \\ 1 \\ 0 \end{bmatrix}, \quad s_3 = \begin{bmatrix} 0 \\ 0 \\ 1 \end{bmatrix}, \quad s_4 = \begin{bmatrix} 0 \\ -1 \\ 1 \end{bmatrix}, \quad s_5 = \begin{bmatrix} -1 \\ 0 \\ 1 \end{bmatrix}$$

In this case, one finds another family of naturally independent sets:

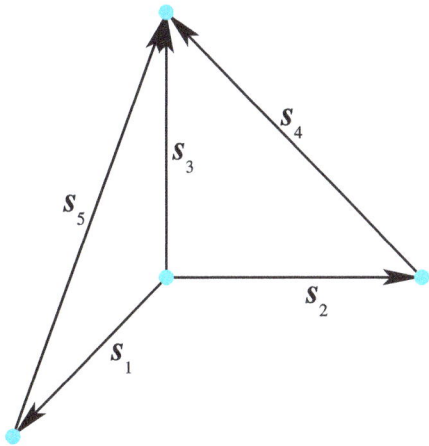

Figure 5.1: A simple Kirchhoff graph for S in Example 5.2 that sets nicely in the corner of the first octant in $\mathbb{R}^3$.

$$\mathcal{I} = \{\emptyset\} \cup \{\text{all singleton subsets of } S\} \cup \{\text{all paired subsets of } S\}$$
$$\cup \{\{s_1, s_2, s_3\}, \{s_1, s_2, s_4\}, \{s_1, s_2, s_5\}, \{s_1, s_3, s_4\},$$
$$\{s_1, s_4, s_5\}, \{s_2, s_3, s_5\}, \{s_2, s_4, s_5\}, \{s_3, s_4, s_5\}\}$$

Two combinations of three vectors that are *not* listed above, $\{s_2, s_3, s_4\}$ and $\{s_1, s_3, s_5\}$, form cycles in the corresponding Kirchhoff graph. An example of a Kirchhoff graph for these edge vectors is shown in Figure 5.1, but keep in mind that any ground set or any vector graph with exactly the same set of dependencies corresponds to the same matroid and the same Kirchhoff graph. Again, it is relatively easy to check that this ground set S and family of independent sets $\mathcal{I}$ satisfy the three conditions, making $(S, \mathcal{I})$ a matroid.

Let A be any $m \times n$ matrix with entries in some field $\mathbb{F}$. Such matrices can be formed using as its columns the vectors of S from either of the previous two examples, and then the matroids discussed in those examples can be associated with those matrices. Also, let G be any (standard) graph, and let $E(G)$ be the edge set of that graph.

Definition. The **column matroid** for the matrix A, denoted $M[A]$, is the matroid whose ground set S is the set of columns of A and whose independent sets $\mathcal{I}$ are all subsets of S that are linearly independent as sets of vectors. The **cycle matroid** for the graph G, denoted $M(G)$, is obtained by taking the edge set $E(G)$ as the ground set S and the family of all acyclic subgraphs of G as the independent sets:

$$\mathcal{I} = \{X \subset E(G) \mid G[X] \text{ is a forest}\}$$

Again, one can check that both $M[A]$ and $M(G)$ satisfy the three conditions in the definition of matroid.

Definition. A matroid M with ground set S having n elements is **representable** over a field $\mathbb{F}$ if there exists an $m \times n$ matrix A whose entries come from $\mathbb{F}$ such that M is isomorphic to the column matroid $M[A]$. Such a matroid is also called $\mathbb{F}$-**representable**, and a **binary matroid** is a representable matroid over $\mathbb{Z}_2$. A matroid M is **graphic** if there exists some graph G such that M is isomorphic to the cycle matroid $M(G)$.

Based on our discussion so far, one might suspect that all matroids are associated with graphs and/or matrices. This is not the case as the following examples make clear: Not every matroid is binary, and not every binary matroid is graphic.

Example 5.3. The *uniform matroid* U_n^r has a ground set S with n elements where a subset of the elements is independent (thus in $\mathcal{I}$) if and only if it contains at most r elements. The matroid discussed in Example 5.1 is isomorphic to U_4^2: there are four elements in the set; any sets of two or fewer elements are independent; any sets of three or more elements contain circuits. Nevertheless, U_4^2 is *not* a binary matroid because it cannot be represented by the columns of a binary matrix. If the columns have only two entries, then there are at most three possible columns, not the needed four:

$$\begin{bmatrix} 1 & 0 & 1 \\ 0 & 1 & 1 \end{bmatrix}$$

On the other hand, if there are three or more entries, then some sets of three columns will be independent.

Example 5.4. Consider the standard binary matrix representation of the Fano plane F_7:

$$A_{F_7} = \begin{array}{c} \begin{array}{ccccccc} s_1 & s_2 & s_3 & s_4 & s_5 & s_6 & s_7 \end{array} \\ \begin{bmatrix} 1 & 0 & 0 & 1 & 1 & 0 & 1 \\ 0 & 1 & 0 & 1 & 0 & 1 & 1 \\ 0 & 0 & 1 & 0 & 1 & 1 & 1 \end{bmatrix} \end{array} \tag{5.1}$$

Of the thirty-five possible combinations of triples of these seven column vectors, there are seven dependent circuits:

$$\mathcal{C} = \{\{s_1, s_2, s_4\}, \{s_1, s_3, s_5\}, \{s_2, s_3, s_6\}, \{s_4, s_5, s_6\}, \{s_1, s_6, s_7\}, \{s_2, s_5, s_7\}, \{s_3, s_4, s_7\}\}$$

The Fano plane (or Fano matroid) is thus binary, but it is famously not graphic: It is not possible to construct a graph whose cycles correspond to the circuits of this matroid; the edges simply cannot fit together the way they would need to form a graph. As will be made clear in Section 5.4, the Fano plane is Kirchhoff graphic.

After this brief and limited introduction to matroids, let us move to Cayley color graphs, Kirchhoff graphs over finite fields and how these relate to matroids.

5.2 Cayley color graphs

One of the key concepts in constructing Kirchhoff graphs over finite fields, including those corresponding to matroids, is the *Cayley color graph*. Let Γ be any group written additively, and let H be any subset of the elements of Γ.

Definition. The **Cayley color graph** (Γ, H) is a directed graph (digraph) with vertex set Γ and edge set $E = \{(g, g + h) : g \in \Gamma, h \in H\}$, that is, (Γ, H) has one vertex associated with each group element of Γ, and every edge is colored (or labeled) by the difference $h = (g + h) - g$.

For each $h \in H$, every vertex is incident to exactly two h edges, one entering and one exiting. The Cayley color graph is finite if and only if the group Γ is finite, and connected if and only if H is a set of generators of Γ.

Given a rational-valued matrix, it is always easy to construct an *infinite* object with *some* of the desired Kirchhoff-graph properties. Consider any matrix in $\mathbb{Q}^{m \times n}$; let A be this matrix multiplied by the least common multiple of the denominators of its entries. Let $s_1, s_2, \ldots, s_n$ be the column vectors of A. The *Cayley vector color graph*

$(\mathbb{Z}^m, \{s_1, s_2, \ldots, s_n\})$ (for simplicity, also referred to as the Cayley color graph) is itself an infinite vector graph (vector array) with edge vectors satisfying the desired cycle condition. Moreover, each vertex is incident to exactly one incoming and one outgoing copy of the edge vector s_i for each i, that is, every vertex v is a null vertex with $\lambda(v) = \mathbf{0}$, and thus the desired orthogonality is trivially satisfied. Such an infinite vector graph, however, is not very revealing for modeling actual networks, and hence the requirement that a Kirchhoff graph be derived from a finite vector graph.

The present discussion does not attempt to study Cayley color graphs in general, but rather considers those related to Kirchhoff graphs. Thus only two additive groups Γ will be considered here: $\mathbb{Z}^m$ and $\mathbb{Z}_p^m$, where $m \in \mathbb{Z}^+$ and p is prime. Examples of infinite and finite Cayley color graphs are given in Example 5.5 and Example 5.6. For the most part, however, the current discussion focusses on finite subgraphs of Cayley color graphs. For a deeper understanding of Cayley color graphs, see, for example, Cayley [3] (1878), Sabiduss [28] (1964), Magnus et al. [16] (1966), or Pisanski and Servatius [20] (2013).

Example 5.5. Let A be an $m \times n$ matrix with entries in $\mathbb{Z}$, and let $s_1, s_2, \ldots, s_n \in \mathbb{Z}^m$ denote the column vectors of A (in this chapter, m is *not* edge multiplicity). The Cayley color graph $(\mathbb{Z}^m, \{s_1, s_2, \ldots, s_n\})$ is an infinite vector graph with edge vectors $\{s_1, s_2, \ldots, s_n\}$ that both exit from and enter into each element (vertex) in $\mathbb{Z}^m$. This graph is connected if $\{s_1, s_2, \ldots, s_n\}$ span $\mathbb{Z}^m$.

Example 5.6. Let $\mathbb{Z}_p$ be the finite field of integers mod p, let $\Gamma = \mathbb{Z}_p^m$, and let H be a set of vectors in $\mathbb{Z}_p^m$. The Cayley color graph (Γ, H) is finite, with p^m vertices and $|H|p^m$ edges (counting multiplicity). Again, each vector in H both exits from and enters into each vertex in Γ, and the graph is connected if and only if the vectors in H span $\mathbb{Z}_p^m$. These graphs and their subgraphs will be considered in Section 5.3.

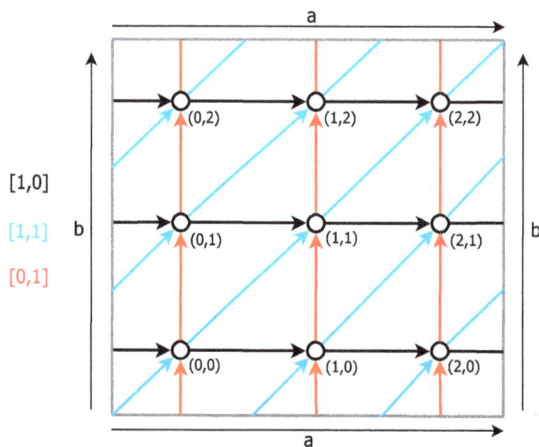

Figure 5.2: The Cayley color graph $(\mathbb{Z}_3^2, \{[1, 0], [1, 1], [0, 1]\})$, drawn on a flat torus.

When $p = 3$, $m = 2$, and $H = \{[1,0], [1,1], [0,1]\}$, this Cayley color graph is shown in Figure 5.2; it has nine vertices and twenty-seven edge vectors. Where appropriate, the vectors in H are identified with the vertices in Γ: as a vector in H, $[1,0] \cong (1,0) \in \Gamma$. Including Figure 5.2, this section will present four vector graphs over $\mathbb{Z}_3^2$ (see Figures 5.3, 5.7, and 5.8); each will be drawn on the flat torus.

5.2.1 Kirchhoff graphs as finite subgraphs of Cayley color graphs over $\mathbb{Z}^m$

Cayley color graphs are not themselves Kirchhoff graphs when Γ is infinite, and they are only trivially Kirchhoff graphs when Γ is a finite field, but they can be used as a foundation for Kirchhoff graphs. For the moment, let $\Gamma = \mathbb{Z}^m$.

Definition. A **vector graph** for the matrix $A := [s_1|s_2|\cdots|s_n]$ is a finite subgraph G of the infinite Cayley vector color graph $(\mathbb{Z}^m, \{s_1, s_2, \ldots, s_n\})$, possibly with multiple copies of an edge vector connecting adjacent vertices.

Thus a vector graph for A has a finite number of vertices and edge vectors, each of which is a column vector of matrix A. It is now possible to give a definition for *Kirchhoff graph* in terms of these Cayley color graphs. Let G be some vector graph for matrix A. Edge vectors may be traversed in either direction regardless of orientation.

Definition. For each cycle C, the **cycle vector**, denoted $\chi(C)$, has entries indexed by $s_1, s_2, \ldots, s_n$. For each i, the ith entry of $\chi(C)$ is the net number of times cycle C traverses an s_i edge. Specifically, add 1 to the ith entry of $\chi(C)$ each time C traverses an s_i vector in the direction of its orientation and subtract 1 for each s_i that is traversed opposite to its orientation.

Proposition 5.2.1. *For any cycle C of vector graph G, $\chi(C) \in \mathrm{Null}(A)$.*

Proof. The proof is straightforward given our definitions for *vector graph* and *cycle vector*. The entries of $\chi(C)$ are the coefficients of vectors traversed in C. As C is a cycle, the corresponding sum of vectors must be zero. Therefore the vector of coefficients $\chi(C)$ must lie in $\mathrm{Null}(A)$. $\square$

Proposition 5.2.1 is the correspondence between cycles and null space vectors that is guaranteed by choosing a vector graph for A. It is analogous to the consistency condition when vectors are assigned to the edges of a digraph as in the Introduction. Now the definition for *vertex cut* or *incidence vector* is the same as before:

Definition. For each vertex v of a vector graph, the **vertex cut** (incidence vector), denoted $\lambda(v)$, has entries indexed by $s_1, s_2, \ldots, s_n$. For each i, the ith entry of $\lambda(v)$ is the net number of s_i that exit vertex v. Equivalently, it is the number of s_i edges exiting v minus the number of s_i edges entering.

Now recall that the Kirchhoff current law translated graphically to the vertices corresponding to elements of Row(A). This can be rephrased by saying that $\lambda(v) \in$ Row(A) for all vertices v. This leads to a Kirchhoff-graph definition based on a Cayley color graph:

Definition. Let G be any cyclic vector graph for matrix A based on the Cayley color graph $(\mathbb{Z}^m, \{s_1, s_2, \ldots, s_n\})$. Then G is a **Kirchhoff graph** for A if $\lambda(v) \in$ Row(A) for all vertices v of G, and the cycle vectors of G span Null(A).

From here, a Kirchhoff graph theory can be developed for Cayley color graphs just as was done earlier in this monograph for vector graphs over the rationals. The interesting new direction to consider now is what happens when $\mathbb{Z}^m$ in our Cayley color graph is replaced by $\mathbb{Z}_p^m$, thus replacing an integral domain with a finite field. As we will see, it is possible in this case not only to prove the existence of Kirchhoff graphs, but even to explicitly give nontrivial Kirchhoff graphs. It should be clear already that Kirchhoff graphs combine elements of linear algebra, graph theory, and group theory. The next section explores these combinations in the context of finite fields.

5.3 Kirchhoff graphs over finite fields

For a prime p, let A_p be an $m \times n$ matrix with entries in $\mathbb{Z}_p$. Again, let $s_1, \ldots, s_n$ denote the column vectors of A_p, now in $\mathbb{Z}_p^m$.

Definition. Any graph obtainable from the Cayley color graph $(\mathbb{Z}_p^m, \{s_1, \ldots, s_n\})$ by assigning nonnegative multiplicities to the edge vectors is a $\mathbb{Z}_p$-**vector graph** for A_p.

Observe that finiteness is no longer a requirement in the definition of $\mathbb{Z}_p$-vector graph since $(\mathbb{Z}_p^m, \{s_1, \ldots, s_n\})$ is itself finite. Let G be a $\mathbb{Z}_p$-vector graph for A_p, and let C be any cycle of G (again, cycles may traverse edges regardless of orientation). The cycle vector for C, now denoted $\chi_p(C)$, is defined analogously to the previous case, except that now all entries are integers mod p.

Proposition 5.3.1. *For any cycle C of $\mathbb{Z}_p$-vector graph G, $\chi_p(C) \in$ Null(A_p).*

Similarly, for each vertex v, the definition of the vertex cut (incidence vector), now denoted $\lambda_p(v)$, is analogous to the previous case, with all entries being integers mod p. This allows the concept of Kirchhoff graphs to be extended to $\mathbb{Z}_p$ matrices:

Definition. Let G be any cyclic $\mathbb{Z}_p$-vector graph for matrix A_p. If $\lambda_p(v) \in$ Row(A_p) for all vertices v of G, and the cycle vectors of G span Null(A_p), then G is a $\mathbb{Z}_p$-**Kirchhoff graph** for A_p.

Example 5.7. Consider the following matrix A_3 with entries in $\mathbb{Z}_3$:

$$A_3 = \begin{array}{cccc} {\scriptstyle s_1} & {\scriptstyle s_2} & {\scriptstyle s_3} & {\scriptstyle s_4} \\ \left[\begin{array}{cccc} 1 & 0 & 1 & 2 \\ 0 & 1 & 1 & 1 \end{array}\right] \end{array} \tag{5.2}$$

Figure 5.3 presents a $\mathbb{Z}_3$-Kirchhoff graph for A_3. All vertex cuts (mod 3) lie in Row(A_3), and all cycle vectors (mod 3) lie in Null(A_3).

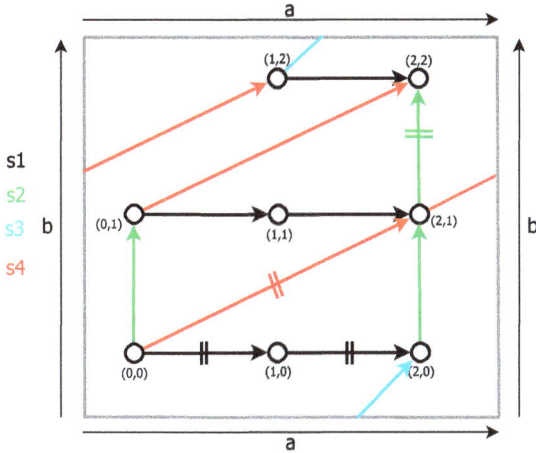

Figure 5.3: A $\mathbb{Z}_3$-Kirchhoff graph for matrix A_3, drawn on the flat torus. Again, hash marks indicate multiplicity; vertex coordinates are given as ordered pairs. Notice that here $m = 1$ (mod 3).

The $\mathbb{Z}_p$-Kirchhoff property also translates to cycle and vertex cut vectors. Let G be a $\mathbb{Z}_p$-Kirchhoff graph for matrix A_p. For any vertex v and any cycle C:

$$\lambda_p(v) \cdot \chi_p(C) \equiv 0 \pmod{p} \tag{5.3}$$

Any vector graph that satisfies (5.3) for all v and C is a $\mathbb{Z}_p$-*Kirchhoff graph*, provided that there are cycles corresponding to a basis for Null(A_p). Note that if a matrix is row-equivalent to A_p (mod p), then a $\mathbb{Z}_p$-Kirchhoff graph for either matrix is a $\mathbb{Z}_p$-Kirchhoff for both matrices.

Interest in studying $\mathbb{Z}_p$-Kirchhoff graphs is motivated by more than simply extending definitions. The primary focus for applications is studying rational-valued matrices, but the rows of any $\mathbb{Q}$-valued matrix can be scaled to give a $\mathbb{Z}$-valued matrix that is row-equivalent. Thus it is sufficient to consider matrices with integer entries. As Lemma 5.3.1 demonstrates, proving the existence (or nonexistence) of Kirchhoff graphs for integer matrices is closely tied to the existence of $\mathbb{Z}_p$-Kirchhoff graphs.

Lemma 5.3.1. *Let A be an integer-valued matrix, and let A_p denote the matrix A (mod p) for any prime p. Let G be a vector graph for A. G is a Kirchhoff graph for A if and only if for all prime p, with all edge vectors taken mod p, it is a $\mathbb{Z}_p$-Kirchhoff graph for A_p.*

Proof. For each p, let G_p be the vector graph obtained by the mod p-reduction of all edge vectors of G. Note that in the process of mod p-reduction, some vertices may be identified. For example, the vertices $(0,0)$ and $(6,3)$ over $\mathbb{Q}$ become the same vertex (mod 3).[1] As G is a vector graph for A, G_p must be a vector graph for A_p, and the cycle vectors of G span $\text{Null}(A)$ if and only if for all primes p, the (mod p) cycle vectors of G_p span $\text{Null}(A_p)$. Finally, G is a Kirchhoff graph if and only if $\lambda(v) \cdot \chi(C) = 0$ for all vertices v and all cycles C. This holds if and only if $\lambda(v) \cdot \chi(C) \equiv 0 \pmod{p}$ for all prime p. This is true, however, if and only if for all vertices v' and all cycles C' of G_p, $\lambda_p(v') \cdot \lambda_p(C') \equiv 0 \pmod{p}$ for all p, that is, if and only if G_p is a $\mathbb{Z}_p$-Kirchhoff graph for all p. □

Although constructing finite Kirchhoff graphs for integer matrices can be difficult, the finite nature of the group $\mathbb{Z}_p^m$ makes the construction of $\mathbb{Z}_p$-Kirchhoff graphs simpler. Indeed, one observation is straightforward:

Proposition 5.3.2. *For a prime p, let A_p be any $m \times n$ matrix with entries in $\mathbb{Z}_p$ and columns $s_1, \ldots, s_n$. Then the Cayley vector color graph $G_{A_p} := (\mathbb{Z}_p^m, \{s_1, \ldots, s_n\})$ is trivially Kirchhoff.*

Proof. G_{A_p} is clearly a $\mathbb{Z}_p$-vector graph for A_p with cycle vectors that span $\text{Null}(A_p)$. As G_{A_p} is a Cayley color graph, every vertex cut is the zero vector, and so G_{A_p} is (trivially) Kirchhoff. □

Although Cayley color graphs over $\mathbb{Z}_p$ are trivially Kirchhoff—in the sense that all vertex cuts are zero—they are the starting point for constructing or showing the existence of graphs that are nontrivially Kirchhoff. Section 5.3.1 will demonstrate that such Kirchhoff graphs exist for many $\mathbb{Z}_p$-valued matrices. Also they are not trivial examples. Example 5.8 shows that a Kirchhoff graph with nonzero vertex cuts, when reduced mod p, can lead to the Cayley color graph as a $\mathbb{Z}_p$-Kirchhoff graph.

Example 5.8. Recall the integer-valued matrix A introduced in Example 5.7:

$$A = \begin{matrix} & \begin{matrix} s_1 & s_2 & s_3 & s_4 \end{matrix} \\ & \begin{bmatrix} 1 & 0 & 1 & 2 \\ 0 & 1 & 1 & 1 \end{bmatrix} \end{matrix} \tag{5.4}$$

The vector graph G in Figure 5.4 is a Kirchhoff graph for this A; indeed, G is a tiling of nine copies of the Kirchhoff graph shown in Figure 5.5. By Lemma 5.3.1, reducing all edge vectors of G mod 3 leads to a $\mathbb{Z}_3$-Kirchhoff graph for $A_3 = A \pmod{3}$. Observe that through mod 3-reduction of edge vectors, some vertices must be identified as well. For example, the vertices $(0,0)$, $(3,0)$, $(3,3)$, and $(6,3)$ in G must become a single vertex in the $\mathbb{Z}_3$-Kirchhoff graph. After mod 3-reduction of all edge vectors of G, the result is precisely the Cayley color graph $(\mathbb{Z}_3^2, \{s_1, \ldots, s_4\})$, with all edges doubled. In particular, although G has nonzero vertex cut vectors, when reduced mod 3, the resulting $\mathbb{Z}_3$-Kirchhoff graph

---

1 A more detailed example of mod p reduction is given in Example 5.8.

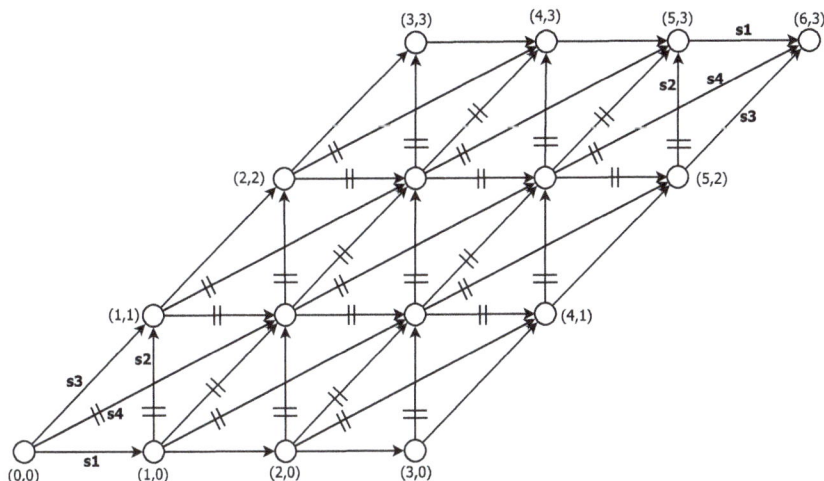

Figure 5.4: G, a tiling of nine copies of the Kirchhoff graph for matrix A in Figure 5.5. When all vectors are reduced mod 3, the result is the Cayley color graph with all edges doubled.

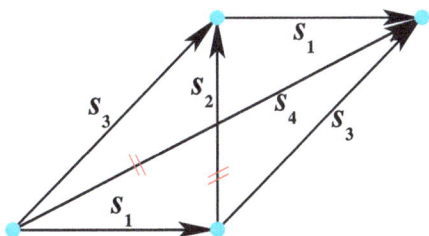

Figure 5.5: A Kirchhoff graph for matrix A in (5.4). As usual, hash marks indicate multiplicity, and vertex coordinates are given by ordered pairs. This is in fact an ordinary Kirchhoff graph for either the rationals or for $\mathbb{Z}_p, p \geq 3$.

has only zero vertex cuts. That is, although Cayley color graphs trivially satisfy the Kirchhoff property, these $\mathbb{Z}_p$-Kirchhoff graphs with only null vertices arise from Kirchhoff graphs with nontrivial vertex cuts.

5.3.1 Nontrivial $\mathbb{Z}_p$-Kirchhoff graphs

This section considers $\mathbb{Z}_p$ matrices that have a $\mathbb{Z}_p$-Kirchhoff graph with at least one nonzero vertex cut.

Definition. A $\mathbb{Z}_p$-Kirchhoff graph is **nontrivial** if it has at least one nonzero vertex cut.

Our next result establishes the existence of nontrivial $\mathbb{Z}_p$-Kirchhoff graphs when p is sufficiently large compared to the matrix dimensions. For any prime p, let A_p be an $m \times n$

matrix with entries in $\mathbb{Z}_p$ and columns $\{s_1, s_2, \ldots, s_n\}$. Note that because row-equivalent matrices have the same $\mathbb{Z}_p$-Kirchhoff graphs, it is sufficient to consider matrices A_p of full row rank.

Theorem 5.3.1. *If $p > n/m$, then A_p has a nontrivial $\mathbb{Z}_p$-Kirchhoff graph.*

Proof. Beginning with the Cayley color graph $G_{A_p} = (\mathbb{Z}_p^m, \{s_1, s_2, \ldots, s_n\})$, assign to each edge vector an undetermined multiplicity (weight), w_i. Since this graph has p^m vertices, and since each vertex in G_{A_p} has n edge vectors exiting it, there are np^m unknown multiplicities. Finding a nontrivial Kirchhoff graph will be equivalent to finding a nontrivial solution to a system of equations in these np^m unknowns. To see that this is possible, recall that as in the linear programming algorithm discussed in Chapter 4, vertex cuts can be written in terms of the multiplicities w_i. For example, a vertex with the vector labels and multiplicities given in Figure 5.6 has the vertex cut:

$$
\begin{array}{ccc}
s_1 & s_2 & s_3 \\
\left[(w_1 - w_2) \right. & (w_3 - w_4) & \left. (w_5 - w_6) \right]
\end{array}
\tag{5.5}
$$

A $\mathbb{Z}_p$-Kirchhoff graph arises whenever $\lambda(v) \cdot \boldsymbol{b} \equiv 0 \pmod{p}$ for all vertices v and all $\boldsymbol{b} \in \text{Null}(A_p)$. As $\text{Null}(A_p)$ has dimension $n - m$, this gives a homogeneous system of $(n - m)p^m$ equations in np^m unknowns. Now we must show that when $mp > n$, there exists a solution that corresponds to a nontrivial $\mathbb{Z}_p$-Kirchhoff graph.

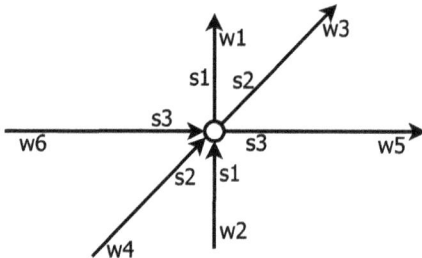

Figure 5.6: An example of edge vectors with unknown multiplicities. The vertex cut for this vertex is $[w_1 - w_2, w_3 - w_4, w_5 - w_6]$.

For each edge vector s_i in G_{A_p}, there is a cycle formed by p copies of s_i and involving no other edge vector. There are p^{m-1} such cycles in G_{A_p}, each of length p. For example, for the edge vector s_2 in Figure 5.7, one such cycle is $v_1 \rightarrow v_4 \rightarrow v_7 \rightarrow v_1$. Assigning multiplicity 1 to each edge in one such cycle and multiplicity 0 to all others is a solution to the system, and the resulting vector graph has all zero vertex cuts. By considering all such cycles for each edge vector s_i there are np^{m-1} such solutions, and they are linearly independent. Moreover, any $\mathbb{Z}_p$-Kirchhoff graph with only zero vertex cuts must assign the same multiplicity to all edges in each such cycle. As a result, any solution of the system that leads to a trivial Kirchhoff graph can be written as a linear combination of

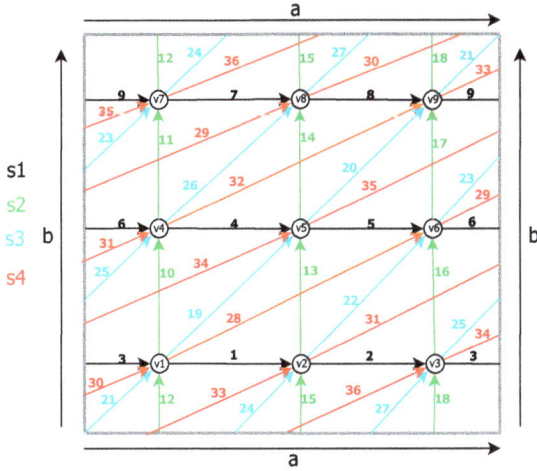

Figure 5.7: Cayley color graph $(\mathbb{Z}_3^2, \{s_1, s_2, s_3, s_4\})$ with as-yet undetermined weights w_i indicated only by their integer subscripts.

these solutions. That is, the system has a np^{m-1}-dimensional solution space that leads to all $\mathbb{Z}_p$-Kirchhoff graphs with only zero vertex cut vectors.

Any solution to the system outside this np^{m-1}-dimensional space gives a set of multiplicities for a nontrivial $\mathbb{Z}_p$-Kirchhoff graph (note the multiplicities can chosen so that $0 \le w_i < p$). Therefore any system whose solution space has dimension greater than np^{m-1} must have a nontrivial Kirchhoff graph. Recalling that the homogeneous system has $(n-m)p^m$ equations in np^m unknowns, the solution space has dimension at least

$$np^m - (n-m)p^m = mp^n = (mp)p^{m-1} > np^{m-1} \tag{5.6}$$

since $mp > n$. Therefore A_p must have a nontrivial Kirchhoff graph. $\qquad\square$

Example 5.9. Consider again the matrix A_3 introduced in Example 5.7:

$$A_3 = \begin{array}{c} \begin{array}{cccc} s_1 & s_2 & s_3 & s_4 \end{array} \\ \left[\begin{array}{cccc} 1 & 0 & 1 & 2 \\ 0 & 1 & 1 & 1 \end{array} \right] \end{array} \tag{5.7}$$

The Cayley color graph $(\mathbb{Z}_3^2, \{s_1, s_2, s_3, s_4\})$ with unknown weights (multiplicities) $w_1, w_2, \ldots, w_{36}$ assigned to each edge is shown in Figure 5.7 (the weights are simply denoted by their integer subscripts). Given that $\{[1\ 1\ -1\ 0]^T, [2\ 1\ 0\ -1]^T\}$ is a basis for Null(A_3), the system of equations can be summarized by the matrix equation $BN = [0]$:

$$
\begin{array}{c}
\begin{array}{cccc}
s_1 & s_2 & s_3 & s_4
\end{array}\\
\begin{array}{c}
\lambda(v_1)\\
\lambda(v_2)\\
\lambda(v_3)\\
\lambda(v_4)\\
\lambda(v_5)\\
\lambda(v_6)\\
\lambda(v_7)\\
\lambda(v_8)\\
\lambda(v_9)
\end{array}
\left[
\begin{array}{cccc}
(w_1 - w_3) & (w_{10} - w_{12}) & (w_{19} - w_{21}) & (w_{28} - w_{30})\\
(w_2 - w_1) & (w_{13} - w_{15}) & (w_{22} - w_{24}) & (w_{31} - w_{33})\\
(w_3 - w_2) & (w_{16} - w_{18}) & (w_{25} - w_{27}) & (w_{34} - w_{36})\\
(w_4 - w_6) & (w_{11} - w_{10}) & (w_{26} - w_{25}) & (w_{32} - w_{31})\\
(w_5 - w_4) & (w_{14} - w_{13}) & (w_{20} - w_{19}) & (w_{35} - w_{34})\\
(w_6 - w_5) & (w_{17} - w_{16}) & (w_{23} - w_{22}) & (w_{29} - w_{28})\\
(w_7 - w_9) & (w_{12} - w_{11}) & (w_{24} - w_{23}) & (w_{36} - w_{35})\\
(w_8 - w_7) & (w_{15} - w_{14}) & (w_{27} - w_{26}) & (w_{30} - w_{29})\\
(w_9 - w_8) & (w_{18} - w_{17}) & (w_{21} - w_{20}) & (w_{33} - w_{32})
\end{array}
\right]
\left[
\begin{array}{cc}
1 & 2\\
1 & 1\\
-1 & 0\\
0 & -1
\end{array}
\right]
$$

$$
=
\begin{bmatrix}
0 & 0\\
0 & 0\\
0 & 0\\
0 & 0
\end{bmatrix}
\ (\mathrm{mod}\ p). \tag{5.8}
$$

Equation (5.8) is the sort of system discussed in the proof of Theorem 5.3.1. Since $m = 2$, $n = 4$, and $p = 3$, implying $p > n/m$, by Theorem 5.3.1, A_3 has a nontrivial $\mathbb{Z}_3$-Kirchhoff graph. One solution to system (5.8) that leads to a nontrivial Kirchhoff graph is

$$
w_1 = w_2 = w_{17} = w_{28} = 2, \quad w_4 = w_5 = w_8 = w_{10} = w_{16} = w_{27} = w_{29} = w_{32} = 1
$$

with all other multiplicities being zero. Removing all edges of multiplicity zero gives the nontrivial $\mathbb{Z}_3$-Kirchhoff graph originally presented in Figure 5.3.

One might wonder if some sort of existence proof along the lines of the proof of Theorem 5.3.1 is possible for Kirchhoff graphs over the rationals. The main difference is that over the rationals, solutions to linear systems with negative weights are possible but not meaningful in terms of Kirchhoff graphs. For finite fields, there are no negatives. This is why a linear algebra problem for finite fields is a linear programming problem for the rationals.

Theorem 5.3.1 guarantees the existence of nontrivial $\mathbb{Z}_p$-Kirchhoff graphs for sufficiently large p, but it does not indicate how such graphs are constructed. The remaining results are constructive. Theorem 5.3.2 shows how to construct a nontrivial $\mathbb{Z}_p$-Kirchhoff graph in many cases, and Theorem 5.4.1 deals with binary matrices and the construction of $\mathbb{Z}_2$-Kirchhoff graphs. Again, for a prime p, let A_p be an $m \times n$ matrix with entries in $\mathbb{Z}_p$ and columns $\{s_1, s_2, \dots, s_n\}$.

Theorem 5.3.2. *If there exists a vector $x \in \mathrm{Row}(A_p)$ with all entries nonzero (mod p), then A_p has a specific nontrivial $\mathbb{Z}_p$-Kirchhoff graph.*

Proof. Let $x_1, x_2, \dots, x_n$ denote the entries of x, where each x_i may be chosen so that $0 < x_i < p$. Beginning with the Cayley color graph $G_{A_p} = (\mathbb{Z}_p^m, \{s_1, s_2, \dots, s_n\})$, the row space vector x can be used to assign multiplicities to each edge vector in G_{A_p}. Choose some vector s_i. As before, the s_i edges of G_{A_p} can be partitioned into p^{m-1} cycles, each of length p. Assign multiplicity x_i to some edge of each cycle. Following edge orientations, traverse each cycle and assign the successive edge vectors multiplicities $2x_i, 3x_i, \ldots, (p-1)x_i, px_i$ respectively. The edge vectors assigned multiplicity px_i may be deleted, and all other multiplicities may be reduced mod p to lie between 1 and $p-1$. Repeating this process for all edge vectors s_i, the result is a vector graph for A_p in which $\lambda_p(v) \equiv x$ (mod p) for all vertices v. As $x \in \mathrm{Row}(A_p)$, it follows that this vector graph is in fact a nontrivial $\mathbb{Z}_p$-Kirchhoff graph for A_p. □

Corollary 5.3.1. *Let A be an integer-valued matrix with no zero columns. Then for sufficiently large prime p, the matrix $A_p = A$ (mod p) has a specific nontrivial $\mathbb{Z}_p$-Kirchhoff graph.*

Remark. Unfortunately, the $\mathbb{Z}_p$-Kirchhoff graph constructed in Corollary 5.3.1 may not be a Kirchhoff graph over the rationals as it may still depend on the finite field and the toroidal embedding.

Example 5.10. Consider another matrix $p = 3$:

$$A_3' = \begin{bmatrix} \overset{s_1}{1} & \overset{s_2}{0} & \overset{s_3}{1} \\ 0 & 1 & 1 \end{bmatrix} \tag{5.9}$$

Summing the two rows, observe that $x = [1\ 1\ 2] \in \mathrm{Row}(A_3')$. Figure 5.8 shows the multiplicities assigned in the proof of Theorem 5.3.2 and the resulting $\mathbb{Z}_3$-Kirchhoff graph.

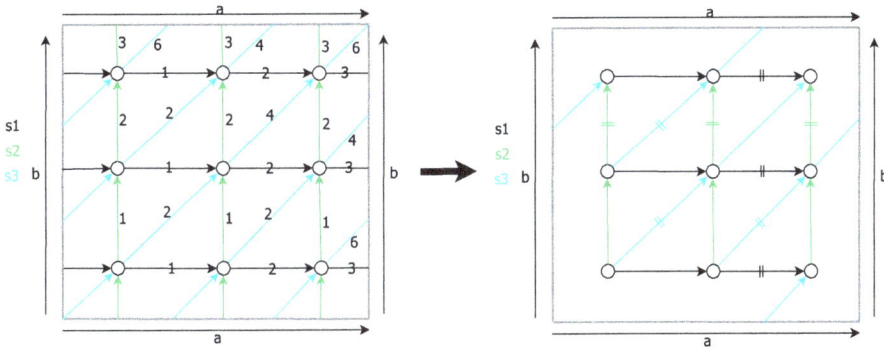

Figure 5.8: Multiplicities assigned as in the proof of Theorem 5.3.2 to the Cayley color graph $(\mathbb{Z}_3^2, \{s_1, s_2, s_3\})$ and the resulting $\mathbb{Z}_3$-Kirchhoff graph for matrix A_3'.

5.4 $\mathbb{Z}_2$-Kirchhoff graphs and binary matroids

Finally, let us turn our attention to binary matroids and binary matrices. The end of this section will discuss some of these results in the context of matroids.

Theorem 5.4.1. *Every nonzero binary matrix has a nontrivial $\mathbb{Z}_2$-Kirchhoff graph.*

Proof. Let A_2 be any $m \times n$ binary matrix with columns $\boldsymbol{s}_1, \boldsymbol{s}_2, \ldots, \boldsymbol{s}_n$, and let $\boldsymbol{G}_{A_2}$ be the Cayley color graph $(\mathbb{Z}_2^m, \{\boldsymbol{s}_1, \boldsymbol{s}_2, \ldots, \boldsymbol{s}_n\})$. As each element of the additive group $\mathbb{Z}_2^m$ is its own inverse, the edges of $\boldsymbol{G}_{A_2}$ occur in pairs: for each edge vector, there is a second copy of that edge vector with the same endpoints and opposite orientation. Let $\boldsymbol{x} = [x_1 \ x_2 \ \cdots \ x_n]$ be any nonzero row of A_2. For each $1 \leq k \leq n$, if $x_k = 1$, then remove one edge vector from every pair of $\boldsymbol{s}_k$ vectors in the graph $\boldsymbol{G}_{A_2}$. The result is a $\mathbb{Z}_2$-vector graph for A_2 with the same cycles as $\boldsymbol{G}_{A_2}$. For all vertices v in the newly constructed graph, however, $\lambda_2(v) \equiv \boldsymbol{x} \pmod 2$, and the result is a nontrivial $\mathbb{Z}_2$-Kirchhoff graph for A_2. $\qquad\square$

Much interest has been given to understanding the relationship between binary matroids and graphs. Given a binary matrix A, there are often many nontrivial $\mathbb{Z}_2$-Kirchhoff graphs for A beyond those constructed in Theorem 5.4.1. For example, suppose the column matroid of matrix A is graphic, that is, isomorphic to the cycle matroid of some graph G. Letting each edge of G be the corresponding (under the matroid isomorphism) column vector of A, the result is a vector graph for A. Since each edge occurs exactly once, the $\mathbb{Z}_2$-Kirchhoff property is now equivalent to the classical orthogonality of the cycle and cut space of graphs. The result is thus a nontrivial $\mathbb{Z}_2$-Kirchhoff graph for A.

On the other hand, many binary matroids—those expressible as the column matroid of a $\mathbb{Z}_2$-valued matrix—are not graphic. A matroid is not graphic if it contains any of five excluded minors: the uniform matroid U_4^2, matroid duals $M^*(K_5)$ and $M^*(K_{3,3})$, and the Fano plane and its dual, F_7 and F_7^* [19]. In addition, Tutte [30–32] developed an algorithm for determining which binary matroids are graphic. This algorithm is constructive: if a matroid is graphic, then the algorithm will produce an appropriate graph. It was recently revisited by Pitsoulis [21] (2014). In the present case, Theorem 5.4.1 has the following corollary that implies that although not every binary matroid is graphic, every binary matroid is Kirchhoff graphic in the sense of $\mathbb{Z}_2$-Kirchhoff graphs.

Corollary 5.4.1. *Every binary matroid has a nontrivial $\mathbb{Z}_2$-Kirchhoff graph.*

Thus although the cycles of classical graphs cannot capture the dependencies of all binary matroids, $\mathbb{Z}_2$-Kirchhoff graphs pose a graphical alternative that can. The cycle vectors now correspond to circuits of the matroid. Moreover, the $\mathbb{Z}_2$-Kirchhoff property guarantees that nonzero vertex cuts correspond to matroid cocircuits.

In the context of matroids, the edge vectors of a $\mathbb{Z}_2$-Kirchhoff graph correspond to the matroid elements. Edge vectors may be repeated in a $\mathbb{Z}_2$-Kirchhoff graph, corresponding to matroid elements appearing more than once. For example, the Fano matroid

F_7 is a binary matroid that is famously not graphic, but as is shown in the next example, it is Kirchhoff graphic.

Example 5.11. The matrix A_{F_7} in (5.10) is the standard binary representation for the Fano plane:

$$A_{F_7} = \begin{array}{c} \begin{array}{ccccccc} s_1 & s_2 & s_3 & s_4 & s_5 & s_6 & s_7 \end{array} \\ \left[\begin{array}{ccccccc} 1 & 0 & 0 & 1 & 1 & 0 & 1 \\ 0 & 1 & 0 & 1 & 0 & 1 & 1 \\ 0 & 0 & 1 & 0 & 1 & 1 & 1 \end{array} \right] \end{array} \qquad (5.10)$$

Figure 5.9 presents two drawings of a nontrivial $\mathbb{Z}_2$-Kirchhoff graph for the matrix A_{F_7}. Note that since $1 \equiv -1 \pmod 2$, edge orientations have been omitted. This graph was not constructed using any of the previous proof techniques. Instead, it is a proper subgraph of the complete Cayley color graph obtained by deleting two vertices (in this case $(1, 0, 1)$ and $(1, 1, 1)$) and some additional edges to maintain the desired vertex conditions. Note that the element s_2 is distinguished as the only doubled edge resulting from the choice of deleted vertices—deleting other pairs of vertices would result in other edge vectors being doubled. Beginning with the complete Cayley color graph, as shown here, six vertices is the minimum number in a proper subgraph that retains all labeled cycles.

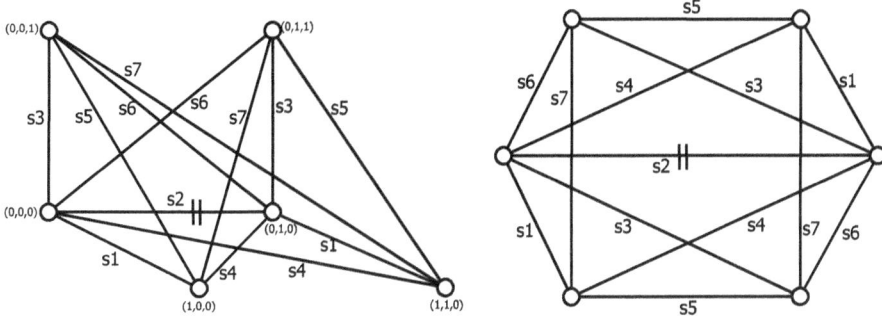

Figure 5.9: Two drawings of a nontrivial $\mathbb{Z}_2$-Kirchhoff graph for the Fano plane, A_{F_7}. The first emphasizes that this graph is a subgraph of the complete Cayley color graph, whereas the second uses symmetry to help indicate which edges are the same vector. Note that since $1 \equiv -1 \pmod 2$, edge orientations are omitted.

Under representation (5.10), the Fano matroid has 14 circuits, 7 of cardinality 3 and 7 of cardinality 4:

$$\mathcal{C}(F_7) = \left\{ \begin{array}{l} \{s_1, s_2, s_4\}, \{s_1, s_3, s_5\}, \{s_1, s_6, s_7\}, \{s_2, s_3, s_6\}, \{s_2, s_5, s_7\}, \\ \{s_3, s_4, s_7\}, \{s_4, s_5, s_6\}, \{s_1, s_4, s_5, s_7\}, \{s_1, s_2, s_3, s_7\}, \{s_1, s_3, s_4, s_6\}, \\ \{s_1, s_2, s_5, s_6\}, \{s_2, s_3, s_4, s_5\}, \{s_2, s_4, s_6, s_7\}, \{s_3, s_5, s_6, s_7\}. \end{array} \right\} \qquad (5.11)$$

Every matroid element appears exactly twice in the $\mathbb{Z}_2$-Kirchhoff graph of Figure 5.9. More importantly, every circuit in (5.11) appears as the edge labels of a cycle. This demonstrates the flexibility afforded by $\mathbb{Z}_2$-Kirchhoff graphs: Although F_7 is not graphic, the cycles of a $\mathbb{Z}_2$-Kirchhoff graph capture the dependencies of F_7 while maintaining vertices that correspond to matroid cocircuits.

5.5 Future considerations: Kirchhoff graph existence over the rationals

The primary open problem in the study of Kirchhoff graphs is the conjecture that every rational-valued matrix has a Kirchhoff graph. This is a difficult question—no general construction method is known. As Lemma 5.3.1 demonstrated, studying $\mathbb{Z}_p$-Kirchhoff graphs could lead to a proof of this conjecture. In particular, Proposition 5.3.2 demonstrated that every $\mathbb{Z}_p$-valued matrix has a $\mathbb{Z}_p$-Kirchhoff graph. This graph—the Cayley color graph—has all zero vertex cuts. Although this structure is somewhat trivial, Example 5.8 demonstrated that a Kirchhoff graph for an integer matrix A, when reduced mod p, leads to the Cayley color graph for the matrix A (mod p). The reverse of this procedure poses an interesting problem: given a matrix A (mod p), can the Cayley color graph be *unfolded* into a finite Kirchhoff graph for A?

Theorem 5.3.1 shows that for any $m \times n$ matrix with entries in $\mathbb{Z}_p$, there exists a nontrivial $\mathbb{Z}_p$-Kirchhoff graph whenever $p > n/m$. The proof of this theorem was of a different nature than most others presented here. As in the linear programming algorithm, an unknown multiplicity was first assigned to each edge vector of the Cayley color graph. Writing each vertex cut in terms of these unknown multiplicities led to a large homogeneous system of linear equations. Finding a nontrivial $\mathbb{Z}_p$-Kirchhoff graph was then equivalent to finding an appropriate solution to this system. Applying a similar approach to rational-valued matrices leads to a linear programming problem since the weights or multiplicities must be nonnegative. The existence question then becomes a question of showing that this linear programming problem has a nontrivial solution. This question can be viewed as trying to find an integer point on a polytope cone in the first hyperoctant of a high-dimensional multiplicity space. Alternatively, based on the dimensions and entries of the matrix, there may be an upper bound on the number of vertices or edges required to guarantee a solution to the system. Such an upper bound could then be used to constrain a computer search algorithm.

6 Equitable Kirchhoff graphs

This chapter considers the connections between Kirchhoff graphs on the one hand and equitable edge partitions on the other. Equitable edge partitions are an extension of the more widely studied concept of equitable vertex partitions (often just called equitable partitions). It is not surprising that Kirchhoff graphs should relate most naturally to equitable edge partitions rather than equitable vertex partitions, since Kirchhoff graphs reflect their edge vector dependencies much as matroids that are based on graphs do for those graphs. Kirchhoff graphs that correspond to equitable edge partitions show an important form of Kirchhoff graph symmetry that is very different from the chiral symmetry discussed previously.

Different types of edge partitioning (other than equitable) have been used over the years: Graph edge partitions are an important research area; they are useful in data decomposition and distributed computing. These methods often involve partitioning the edges of a (large) graph into connected subgraphs to minimize some objective. This chapter, however, only considers equitable edge partitions for multidigraphs.

Section 6.1 discusses equitable edge partitions and their relation to vector graphs. Section 6.2 defines and discusses Kirchhoff edge partitions for directed graphs, making the connection between Kirchhoff graphs and equitable edge partitions. The key result is that if the natural edge partition based on edge vectors (one distinct edge vector per partition class for D) is equitable, and if the quotient matrix based on the edge partition is symmetric, then the vector graph is Kirchhoff. On the other hand, the natural edge partitions for many Kirchhoff graphs, including those containing larger cycles, need not be equitable, and counterexamples disproving such a result are given in Section 6.3. Finally, the relationship between the symmetry of quotient matrices and the uniformity of equitable edge partitions is discussed in Section 6.4.

One might wonder if it is possible to construct a Kirchhoff graph from an equitable natural edge partition and a symmetric matrix serving as the quotient matrix. In principle, this may be possible if the symmetric matrix satisfies the constraints to be a quotient matrix, but still the construction would be nontrivial.

6.1 Connecting vector graphs to equitable edge partitions

To connect vector graphs with equitable edge partitions, a new associated digraph D based on the vector graph G is needed. This *associated multidigraph* is perhaps best described through an example:

Example 6.1. Consider the Kirchhoff graph (which of course is also a vector graph) first presented in the Introduction and shown in Figures 1.11 and 2.4. Figure 2.4a shows this Kirchhoff graph, and its original associated digraph is shown in Figure 2.4b. The *associated multidigraph D* for this Kirchhoff graph, together with the Kirchhoff graph itself, is

https://doi.org/10.1515/9783111408576-006

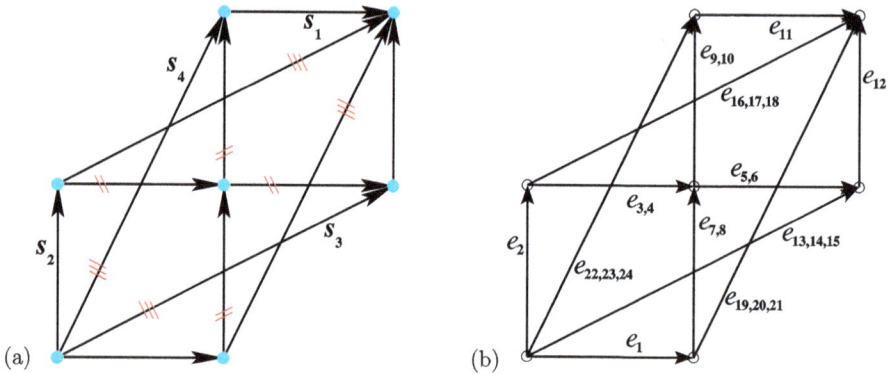

Figure 6.1: A second Kirchhoff-digraph pair. (a) This Kirchhoff graph **G** is generated by any set of vectors $\mathcal{S} = \{s_1, s_2, s_3, s_4\}$ whose dependencies are given by the matrices R and N in (1.10). Hash marks again indicate the number of copies of a given edge vector lying in parallel connecting the same two vertices. Each edge vector is represented by the corresponding column vector of R and occurs uniformly a total of six times. (b) The associated multidigraph having 24 edges, one for each of the six copies of the four edge vectors in $\mathcal{S}$.

shown in Figure 6.1. This second digraph has a distinct edge for each copy of each edge vector in **G**. Thus for this Kirchhoff graph, D is a multidigraph with 24 total edges (six distinct edges for each edge vector s_k, $1 \le k \le 4$). The direction of each edge in D is again the same as the positive direction of the corresponding edge vector in **G**. This new associated digraph D still resembles the vector graph **G** and may even be identical to the original associated digraph if the vector graph has a single edge vector between adjacent vertices. Nonetheless, this new digraph is called the *associated multidigraph*. The distinction between **G** and D is just that instead of having a certain number of copies of an edge vector between two adjacent vertices, D can be a multigraph with that number Tof directed edges between these vertices.

Although much of the subject matter here pertains equally to the equitable edge partitions of any digraph, the natural partitioning of the digraph edges based on their associated edge vectors in a vector graph is particularly important.

Definition. For a vector graph G, the **natural edge partition** for an associated digraph D is the partition where each partition class corresponds to a unique edge vector and contains all the directed edges of D that correspond to that edge vector in **G**.

The next subsection introduces the concept of signed adjacency between directed edges in a digraph D, and this leads to a signed *edge* adjacency matrix for D. Using this matrix, equitable edge partitions are defined, and the classical results for equitable vertex partitions continue to hold, now with the edge adjacency matrix A_E in place of adjacency matrix A.

It is also worth noting that if the negative signs in $A_E(D)$ are omitted, then the result is the adjacency matrix $A_{L(D)}$ for the line graph of the underlying graph D. Harary and Norman [15] (1960) introduced a directed line graph of a directed graph D by putting a directed edge from edge e to edge f if e and f form a directed path in D from the base (initial vertex) of e to the terminus of f. Its adjacency matrix $HN(D)$ is not symmetric, but adding to it its transpose results in a symmetric matrix: $A_E(D) = A_{L(D)} - 2(HN(D) + HN(D)^{\mathsf{T}})$. Up to reversal of all edge directions, a connected D is recoverable from $A_E(D)$.

6.1.1 Edge partitions of digraphs

Let D be a digraph (possibly a multidigraph) with vertices $V(D) = \{v_i\}$ and edges $E(D) = \{e_j\}$.

Definition. Define two functions $\iota, \tau : E(D) \rightarrow V(D)$, where $\iota(e)$ is the **initial vertex**, and $\tau(e)$ is the **terminal vertex**, each for edge e, that is, every edge e is of the form $e = (\iota(e), \tau(e))$. Now for each $i \neq j \in \{1, \ldots, m\}$, define the **edge pair orientation**:

$$o_{i,j} := \begin{cases} 1 & \text{if either } \iota(e_i) = \iota(e_j) \text{ or } \tau(e_i) = \tau(e_j) \\ 2 & \text{if } \iota(e_i) = \iota(e_j) \text{ and } \tau(e_i) = \tau(e_j) \\ -1 & \text{if either } \iota(e_i) = \tau(e_j) \text{ or } \tau(e_i) = \iota(e_j) \\ -2 & \text{if } \iota(e_i) = \tau(e_j) \text{ and } \tau(e_i) = \iota(e_j) \\ 0 & \text{otherwise} \end{cases} \tag{6.1}$$

The **signed edge adjacency matrix** of D, $A_E(D)$, is the symmetric $|E(D)| \times |E(D)|$ matrix with (i, j)th entry $o_{i,j}$.

Thus $|o_{i,j}| = 2$ if e_i and e_j share both endpoints and is negative when they have opposite orientations, and $|o_{i,j}| = 1$ if e_i and e_j share one endpoint and is negative when that vertex is the base vertex of one edge and the terminal vertex of the other. Let $o_{i,i} = 0$ for all i. The values of this function are summarized in Figure 6.2.

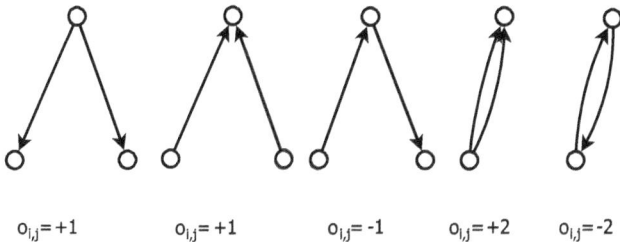

$o_{i,j} = +1$ $o_{i,j} = +1$ $o_{i,j} = -1$ $o_{i,j} = +2$ $o_{i,j} = -2$

Figure 6.2: The five edge configurations for which $o_{i,j} \neq 0$. Notice that it does not matter which directed edge is i and which is j, so the signed edge adjacency matrix $A_E(D)$ is indeed symmetric.

Example 6.2. Figure 6.3 shows a vector (and Kirchhoff) graph along with an associated multidigraph that in this case is just a digraph. This digraph D has a signed edge adjacency matrix:

$$
A_E(D) = \begin{array}{c}
 \\
e_1 \\
e_2 \\
e_3 \\
e_4 \\
e_5 \\
e_6 \\
e_7 \\
e_8
\end{array}
\begin{array}{cccccccc}
e_1 & e_2 & e_3 & e_4 & e_5 & e_6 & e_7 & e_8 \\
\left[\begin{array}{cccccccc}
0 & -1 & 1 & 0 & -1 & 1 & -1 & 0 \\
-1 & 0 & 0 & 1 & 1 & -1 & 0 & -1 \\
1 & 0 & 0 & 0 & 1 & 0 & -1 & 1 \\
0 & 1 & 0 & 0 & 0 & 1 & 1 & -1 \\
-1 & 1 & 1 & 0 & 0 & -1 & 0 & 1 \\
1 & -1 & 0 & 1 & -1 & 0 & 1 & 0 \\
-1 & 0 & -1 & 1 & 0 & 1 & 0 & 0 \\
0 & -1 & 1 & -1 & 1 & 0 & 0 & 0
\end{array}\right]
\end{array}
$$

Notice that for digraphs derived from Kirchhoff graphs where no edge vector in S is a scalar multiple of another edge vector in S, the signed edge adjacency matrix will never have -2 as an entry. Also, the entry 2 only occurs when two directed edges correspond to the same edge vector. This example will be revisited repeatedly throughout this section.

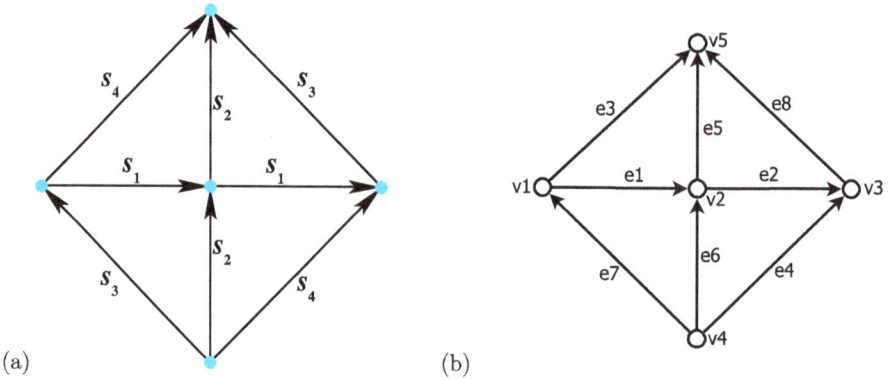

(a) (b)

Figure 6.3: (a) A Kirchhoff graph G with four edge vectors. (b) The associated (multi)digraph D with five vertices and eight edges that correspond to G.

For any digraph D, $A_E(D)$ is a symmetric matrix. Moreover, $A_E(D)$ can be computed from a number of other graph matrices. Let $B = B(D)$ be the *incidence matrix* of digraph D, that is, B is a $|V(D)| \times |E(D)|$ matrix with (i, j)th entry 1 if $v_i = \iota(e_j)$ and -1 if $v_i = \tau(e_j)$. Otherwise, $B_{i,j}$ is 0 when vertex v_i is not an endpoint of edge e_j. For example, the digraph in Figure 6.3 (Example 6.2) has the incidence matrix

$$
B(D) = \begin{array}{c} \\ v_1 \\ v_2 \\ v_3 \\ v_4 \\ v_5 \end{array}
\begin{array}{cccccccc}
e_1 & e_2 & e_3 & e_4 & e_5 & e_6 & e_7 & e_8 \\
\end{array}
\left[\begin{array}{cccccccc}
1 & 0 & 1 & 0 & 0 & 0 & -1 & 0 \\
-1 & 1 & 0 & 0 & 1 & -1 & 0 & 0 \\
0 & -1 & 0 & -1 & 0 & 0 & 0 & 1 \\
0 & 0 & 0 & 1 & 0 & 1 & 1 & 0 \\
0 & 0 & -1 & 0 & -1 & 0 & 0 & -1
\end{array} \right]
$$

Let the incidence vector (vertex cut) $\lambda(v)$ for a vertex v again be the vector with $|E(D)|$ entries, the jth entry 1 if $v = \iota(e_j)$, -1 if $v = \tau(e_j)$, and 0 otherwise. In particular, $\lambda(v_j)$ is the jth row of the incidence matrix $B(D)$. The entries $o_{i,j}$ and therefore the rows of $A_E(D)$ can be written in terms of vertex cuts.

Proposition 6.1.1. *For each $i \neq j$:*

$$o_{i,j} = \lambda(\iota(e_i))_j - \lambda(\tau(e_i))_j \tag{6.2}$$

Proof. For $i \neq j$:

$$
\lambda(\iota(e_i))_j = \begin{cases} 1 & \text{if } \iota(e_i) = \iota(e_j) \\ -1 & \text{if } \iota(e_i) = \tau(e_j) \\ 0 & \text{otherwise} \end{cases}
\quad \text{and} \quad
-\lambda(\tau(e_i))_j = \begin{cases} 1 & \text{if } \tau(e_i) = \tau(e_j) \\ -1 & \text{if } \tau(e_i) = \iota(e_j) \\ 0 & \text{otherwise} \end{cases} \tag{6.3}
$$

Therefore by the definition of $o_{i,j}$ (6.1), equation (6.2) holds for $i \neq j$. □

On the other hand, $\lambda(\iota(e_i))_i = -\lambda(\tau(e_i))_i = 1$ for all i, meaning that

$$\lambda(\iota(e_i))_i - \lambda(\tau(e_i))_i = 2 \neq 0 = o_{i,i} \tag{6.4}$$

so the value is different in this case.

Let $V_{\mathcal{I}}$ and $V_{\mathcal{T}}$ be $|E(D)| \times |E(D)|$ matrices with jth row $\lambda(\iota(e_j))$ and $\lambda(\tau(e_j))$, respectively, that is, $V_{\mathcal{I}}$ is a square matrix whose jth row is the incidence vector of the *initial* vertex of edge e_j (hence the subscript $\mathcal{I}$ for "initial"). Similarly, the jth row of the matrix $V_{\mathcal{T}}$ is the incidence vector of the *terminal* vertex of edge e_j ($\mathcal{T}$ for "terminal"). Proposition 6.1.1 and (6.4) together prove the following result.

Corollary 6.1.1.

$$A_E(D) = V_{\mathcal{I}} - V_{\mathcal{T}} - 2I$$

Remark. Although the signed edge adjacency matrix A_E captures interactions between the edges of a digraph, observe that reconstructing a digraph from $A_E(D)$ is nontrivial. However, up to reversal of all edge directions, a connected D is recoverable from $A_E(D)$.

6.1.2 Equitable edge partitions

Again, let D be any digraph with signed edge adjacency matrix $A_E(D)$.

Definition. Let $\pi := \{E_1, \ldots, E_k\}$ be any partition of $E(D)$. A partition π is an **equitable edge partition** if for all $i, j \in \{1, \ldots, k\}$ and all edges $e_p \in E_i$, the number

$$d_{i,j} = \sum_{q : e_q \in E_j} A_E(D)_{p,q} \tag{6.5}$$

depends only on i and j, and not on the choice of edge $e_p \in E_i$.

Remarks.

1. Notice that this sum is taken across the entries in a row of $A_E(D)$ for columns corresponding to the elements of a fixed partition class E_j. *Equitable* means that this sum is the same regardless of which row corresponding to an element of the partition class E_i is chosen.

2. Observe that this definition of an equitable edge partition is analogous to the standard definition of an equitable vertex partition with the signed edge adjacency matrix in (6.5) taking the place of the adjacency matrix. For an introduction to equitable vertex partitions, see, for example, Godsil [13, Chapter 5] (1993).

For Example 6.2, the digraph D in Figure 6.3 has a natural edge partition based on the Kirchhoff graph in the same figure: $\pi = \{E_1, E_2, E_3, E_4\}$, where

$$E_1 = \{e_1, e_2\}, \quad E_2 = \{e_3, e_4\}, \quad E_3 = \{e_5, e_6\}, \quad E_4 = \{e_7, e_8\}$$

are the four partition classes. This edge partition is equitable and is illustrated with colored edges in Figure 6.4.

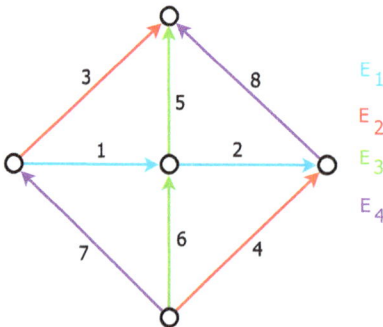

Figure 6.4: A digraph D with an equitable edge partition π with four classes. Numbers k label the directed edges e_k. Directed edges of the same color are in the same equivalence set and are assigned the same vector in the corresponding Kirchhoff graph. When D is viewed as the associated digraph for the vector graph G in Figure 6.3a, this is the natural edge partition.

Remark. Every digraph has an edge partition that is trivially equitable, that is, the partition for which every cell contains exactly one edge.

Definition. Let $A_E(D/\pi)$ be the $k \times k$ matrix with (i,j)th entry d_{ij}. This matrix is called the **quotient matrix** of D with respect to π.

For Example 6.2:

$$A_E(D/\pi) = \begin{array}{c} E_1 \\ E_2 \\ E_3 \\ E_4 \end{array} \begin{array}{cccc} E_1 & E_2 & E_3 & E_4 \\ \left[\begin{array}{cccc} -1 & 1 & 0 & -1 \\ 1 & 0 & 1 & 0 \\ 0 & 1 & -1 & 1 \\ -1 & 0 & 1 & 0 \end{array}\right] \end{array}$$

Remark. Just as the matrix $A_E(D)$ encodes how some edge interacts with the other edges of a digraph, so the matrix $A_E(D/\pi)$ encodes how any edge of a particular partition class interacts with the other partition classes. For example, under the coloring in Figure 6.4, $A_E(D/\pi)_{1,2} = 1$ indicates that every blue edge meets one red edge, the orientations of those edges match at the incident vertex, and no red and blue edges meet at any vertex with opposite orientations. On the other hand, $A_E(D/\pi)_{2,4} = 0$ because red and purple edges meet in D twice with the same orientation and twice with the opposite orientations.

The quotient matrix $A_E(D/\pi)$ is closely related to the signed edge adjacency matrix $A_E(D)$. Let T be the characteristic matrix of edge partition π, that is, T is an $|E(D)| \times k$ matrix with $T_{i,j} = 1$ when $e_i \in E_j$ and $T_{i,j} = 0$ otherwise. For Example 6.2:

$$T = \begin{array}{c} e_1 \\ e_2 \\ e_3 \\ e_4 \\ e_5 \\ e_6 \\ e_7 \\ e_8 \end{array} \begin{array}{cccc} E_1 & E_2 & E_3 & E_4 \\ \left[\begin{array}{cccc} 1 & 0 & 0 & 0 \\ 1 & 0 & 0 & 0 \\ 0 & 1 & 0 & 0 \\ 0 & 1 & 0 & 0 \\ 0 & 0 & 1 & 0 \\ 0 & 0 & 1 & 0 \\ 0 & 0 & 0 & 1 \\ 0 & 0 & 0 & 1 \end{array}\right] \end{array}$$

The matrix T provides a relationship between $A_E(D/\pi)$ and $A_E(D)$. Godsil and McKay [14] (1980) proved a general result about equitable partitions of matrices. The following proposition and its proof are included here for completeness and development of matrix relationships.

Proposition 6.1.2. *Let π be a partition of $E(D)$ with characteristic matrix T. Then π is an equitable edge partition of D if and only if:*

$$A_E(D)T = TA_E(D/\pi) \tag{6.6}$$

Proof. ($\Rightarrow$) For any $p \in \{1,\ldots,|E(D)|\}$, suppose that $e_p \in E_i$, and consider $[A_E(D)T]_{p,j}$. Since for all q, $T_{q,j} = 1$ if and only if $e_q \in E_j$, then

$$[A_E(D)T]_{p,j} = \sum_{q=1}^{|E(D)|} A_E(D)_{p,q} T_{q,j} = \sum_{q:e_q \in E_j} A_E(D)_{p,q} = d_{i,j}$$

by (6.5). On the other hand, as $e_p \in E_i$, row p of T is zero except for a 1 in column i. Therefore row p of $TA_E(D/\pi)$ is row i of $A_E(D/\pi)$, and $[TA_E(D/\pi)]_{p,j} = A_E(D/\pi)_{i,j} = d_{i,j}$. Therefore $A_E(D)T = TA_E(D/\pi)$.

($\Leftarrow$) As $A_E(D)T = TA_E(D/\pi)$, the columns of $A_E(D)T$ are linear combinations of the columns of T and therefore constant on the cells of π, that is, if e_p and e_q belong to class E_j of edge partition π, then:

$$[A_E(D)T]_{p,j} = [A_E(D)T]_{q,j} \tag{6.7}$$

However, for all i and j,

$$[A_E(D)T]_{i,j} = \sum_{q=1}^{|E(D)|} A_E(D)_{i,q} T_{q,j} = \sum_{q:e_q \in E_j} A_E(D)_{i,q}$$

By (6.7), however, this last sum depends only on the partition class containing edge e_i and not on the choice of edge. Therefore by definition, π is an equitable edge partition. $\square$

The quotient matrix $A_E(D/\pi)$ can also be written in terms of vertex cuts and the characteristic matrix T. For each partition class $E_1,\ldots,E_k$, choose some representative edge $\varepsilon_i \in E_i$. Let $\tilde{V}_I$ be a $k \times |E(D)|$ matrix whose ith row is $\lambda(\iota(\varepsilon_i))$, that is, the incidence vector of the *initial* vertex of ε_i. Similarly, let $\tilde{V}_T$ be a $k \times |E(D)|$ matrix whose ith row is $\lambda(\tau(\varepsilon_i))$, the incidence vector of the *terminal* vertex of ε_i.

Proposition 6.1.3.

$$A_E(D/\pi) = (\tilde{V}_I - \tilde{V}_T)T - 2I_k \tag{6.8}$$

Proof. For any $i \neq j \in \{1,\ldots,k\}$, suppose, without loss of generality, that $e_p \in E_i$ is the edge chosen for row i of $\tilde{V}_I$ and $\tilde{V}_T$. Then

$$[(\tilde{V}_I - \tilde{V}_T)T]_{i,j} = \sum_{q=1}^{|E(D)|} [\lambda(\iota(e_p))_q - \lambda(\tau(e_p))_q] T_{q,j}$$

$$= \sum_{q=1}^{|E(D)|} o_{p,q} T_{q,j} \quad \text{by Proposition 6.1.1}$$

$$= \sum_{q:e_q \in E_j} o_{p,q} = \sum_{q:e_q \in E_j} A_E(D)_{p,q}$$

$$= d_{i,j} = A_E(D/\pi)_{i,j}$$

As $I_{i,j} = 0$ for $i \neq j$, it follows that (6.8) holds for all (i,j)-entries when $i \neq j$. A similar argument shows that $[(\tilde{V}_{\mathcal{I}} - \tilde{V}_{\mathcal{I}})T]_{j,j} = d_{j,j} + 2$, and thus (6.8) holds. □

Corollary 6.1.2. *If π is an equitable edge partition, then:*

$$(V_{\mathcal{I}} - V_{\mathcal{T}})T = T(\tilde{V}_{\mathcal{I}} - \tilde{V}_{\mathcal{T}})T \tag{6.9}$$

Proof. Following the previous results:

$$
\begin{aligned}
(V_{\mathcal{I}} - V_{\mathcal{T}})T &= A_E(D)T + 2T && \text{(Corollary 6.1.1)} \\
&= TA_E(D/\pi) + 2T && \text{(Proposition 6.1.2)} \\
&= T((\tilde{V}_{\mathcal{I}} - \tilde{V}_{\mathcal{T}})T - 2I_k) + 2T && \text{(Proposition 6.1.3)} \\
&= T(\tilde{V}_{\mathcal{I}} - \tilde{V}_{\mathcal{T}})T && \square
\end{aligned}
$$

6.2 Connecting equitable edge partitions to Kirchhoff graphs

This section ties the equitable edge partitions of the previous section to Kirchhoff graphs. Let D be any digraph that may have multiple edges and that may or may not have been derived from a vector graph. For a cycle C in a digraph D, let $\chi(C)$ be the cycle vector for C in D. Our previous work suggests a natural definition of a Kirchhoff edge partition of D that matches what it means for a vector graph to be a Kirchhoff graph:

Definition. An edge partition $\pi = \{E_1, \ldots, E_k\}$ of $E(D)$ with characteristic matrix T is a **Kirchhoff edge partition** (or just a **Kirchhoff partition**) if for all vertices v and cycles C:

$$\lambda(v)T \cdot \chi(C)T = 0 \tag{6.10}$$

Regarding the vectors $\lambda(v)T$ and $\chi(C)T$, each has k entries, indexed by $E_1, \ldots, E_k$. The ith entry of the vector $\lambda(v)T$ is the net number of edges of class E_i that exit vertex v. Similarly, the ith entry of vector $\chi(C)T$ is the net number of edges of class E_i that cycle C traverses with the correct orientation. If D is the associated multidigraph of a vector graph G, then for the natural edge partition, this condition exactly matches that in the standard definition for G being Kirchhoff.

Theorem 6.2.1. *Suppose that D is the associated multidigraph of a cyclic vector graph G, and suppose that the natural edge partition of the edge vectors in G is a Kirchhoff partition. Then G is a Kirchhoff graph.*

Every digraph has an edge partition that is Kirchhoff, the trivial edge partition. In this case, the characteristic matrix T is simply the $|E(D)| \times |E(D)|$ identity matrix. Therefore (6.10) becomes

$$\lambda(v) \cdot \chi(C) = 0,$$

which is the classical orthogonality of the cut and cycle spaces of digraphs. The interesting Kirchhoff edge partitions, of course, are the nontrivial Kirchhoff edge partitions.

Again for the example shown in Figure 6.4, its equitable edge partition is also Kirchhoff. Specifically, for any vertex v of D:

$$\lambda(v)T \in \{\pm[0\ \ 1\ \ 1\ \ 1], \pm[1\ \ 1\ \ 0\ \ {-1}], [0\ \ 0\ \ 0\ \ 0]\} \tag{6.11}$$

while for any cycle C:

$$\chi(C)T \in \text{span}\{[1\ \ {-1}\ \ 1\ \ 0], [1\ \ 0\ \ {-1}\ \ 1]\} \tag{6.12}$$

As every vector in (6.11) is orthogonal to every vector in (6.12), the edge partition π is Kirchhoff.

Theorem 6.2.2 will demonstrate that finding Kirchhoff partitions is closely related to finding equitable edge partitions. Let D be any digraph with incidence matrix B, and let $\pi = \{E_1, \ldots, E_k\}$ be an edge partition of D with characteristic matrix T.

Proposition 6.2.1. *Edge partition $\pi = \{E_1, \ldots, E_k\}$ of $E(D)$ is Kirchhoff if and only if the row space of B is invariant under right multiplication by TT^{T}.*

Proof. Let N be a matrix with $|E(D)|$ rows whose columns are a basis for Null(B). As every cycle vector of D lies in the null space of B—and thus the row space of N^{T}—π is Kirchhoff if and only if $BT(N^{\mathsf{T}}T)^{\mathsf{T}} = 0$, that is, if and only if

$$B(TT^{\mathsf{T}})N = 0 \tag{6.13}$$

However, (6.13) holds if and only if the rows of $B(TT^{\mathsf{T}})$ are orthogonal to the columns of N, that is, if and only if the row space of B is invariant under right multiplication by TT^{T}. ☐

Theorem 6.2.2. *Let D be a digraph with equitable edge partition $\pi = \{E_1, \ldots, E_k\}$. If the quotient matrix $A_E(D/\pi)$ is symmetric, then π is a Kirchhoff partition.*

Proof. Let the matrices $V_{\mathcal{I}}, V_{\mathcal{T}}, \tilde{V}_{\mathcal{I}}, \tilde{V}_{\mathcal{T}}$ be defined as before. For any D, the matrix $A_E(D)$ is symmetric, and thus by Corollary 6.1.1 the matrix $(V_{\mathcal{I}} - V_{\mathcal{T}})$ is symmetric as well. By assumption $A_E(D/\pi)$ is symmetric, so by Proposition 6.1.3, $(\tilde{V}_{\mathcal{I}} - \tilde{V}_{\mathcal{T}})T$ is also symmetric. Since π is equitable, by Corollary 6.1.2:

$$(V_{\mathcal{I}} - V_{\mathcal{T}})T = T(\tilde{V}_{\mathcal{I}} - \tilde{V}_{\mathcal{T}})T \tag{6.14}$$

Taking the transpose of (6.14) produces:

$$T^{\mathsf{T}}(V_{\mathcal{I}} - V_{\mathcal{T}})^{\mathsf{T}} = T^{\mathsf{T}}(\tilde{V}_{\mathcal{I}} - \tilde{V}_{\mathcal{T}})^{\mathsf{T}}T^{\mathsf{T}}$$
$$= [(\tilde{V}_{\mathcal{I}} - \tilde{V}_{\mathcal{T}})T]^{\mathsf{T}}T^{\mathsf{T}}$$

Since both $V_{\mathcal{I}} - V_{\mathcal{T}}$ and $(\tilde{V}_{\mathcal{I}} - \tilde{V}_{\mathcal{T}})T$ are symmetric:

$$T^{\mathrm{T}}(V_{\mathcal{I}} - V_{\mathcal{T}}) = (\tilde{V}_{\mathcal{I}} - \tilde{V}_{\mathcal{T}})TT^{\mathrm{T}}$$

Therefore

$$TT^{\mathrm{T}}(V_{\mathcal{I}} - V_{\mathcal{T}}) = T(\tilde{V}_{\mathcal{I}} - \tilde{V}_{\mathcal{T}})TT^{\mathrm{T}}$$

and by (6.14): $\qquad\qquad TT^{\mathrm{T}}(V_{\mathcal{I}} - V_{\mathcal{T}}) = (V_{\mathcal{I}} - V_{\mathcal{T}})TT^{\mathrm{T}}$ $\qquad\qquad$ (6.15)

However, the rows of $(V_{\mathcal{I}} - V_{\mathcal{T}})$ span the row space of Q. Therefore the rows of $TT^{\mathrm{T}}(V_{\mathcal{I}} - V_{\mathcal{T}})$ all lie in the row space of Q, and so by (6.15) each row of $(V_{\mathcal{I}} - V_{\mathcal{T}})TT^{\mathrm{T}}$ lies in Row(Q) as well. Thus the row space of Q is invariant under right multiplication by TT^{T}, and by Proposition 6.2.1, π is Kirchhoff. $\qquad\square$

Consider again the Kirchhoff graph G in Figure 6.1a and its associated multidigraph D in Figure 6.1b. Notice that the first six directed edges of D correspond to s_1, the next six correspond to s_2, and so forth. This digraph D has the signed edge adjacency matrix:

$A_E(D) =$

	e_1	e_2	e_3	e_4	e_5	e_6	e_7	e_8	e_9	e_{10}	e_{11}	e_{12}	e_{13}	e_{14}	e_{15}	e_{16}	e_{17}	e_{18}	e_{19}	e_{20}	e_{21}	e_{22}	e_{23}	e_{24}
e_1	0	2	-1	-1	0	0	-1	1	1	-1	-1	0	0	0	0	1	1	1	0	0	0	0	0	0
e_2	2	0	-1	-1	0	0	-1	1	1	-1	-1	0	0	0	0	1	1	1	0	0	0	0	0	0
e_3	-1	-1	0	2	0	0	0	-1	-1	1	1	-1	1	1	1	0	0	0	0	0	0	0	0	0
e_4	-1	-1	2	0	0	0	0	-1	-1	1	1	-1	1	1	1	0	0	0	0	0	0	0	0	0
e_5	0	0	0	0	0	0	-1	-1	1	0	0	0	0	0	0	1	1	1	1	1	1	-1	-1	-1
e_6	0	0	0	0	0	0	1	-1	-1	0	0	0	1	1	1	0	0	0	-1	-1	-1	1	1	1
e_7	-1	-1	0	0	0	1	0	0	0	0	0	0	1	1	1	-1	-1	-1	0	0	0	1	1	1
e_8	1	1	-1	-1	0	-1	0	0	2	-1	-1	0	0	0	0	0	0	0	1	1	1	0	0	0
e_9	1	1	-1	-1	0	-1	0	2	0	-1	-1	0	0	0	0	0	0	0	1	1	1	0	0	0
e_{10}	-1	-1	1	1	-1	0	0	-1	-1	0	2	0	0	0	0	0	0	0	0	0	0	1	1	1
e_{11}	-1	-1	1	1	-1	0	0	-1	-1	2	0	0	0	0	0	0	0	0	0	0	0	1	1	1
e_{12}	0	0	-1	-1	1	0	0	0	0	0	0	0	-1	-1	-1	1	1	1	1	1	1	0	0	0
e_{13}	0	0	1	1	0	1	1	0	0	0	0	-1	0	2	2	0	0	0	0	0	0	1	1	1
e_{14}	0	0	1	1	0	1	1	0	0	0	0	-1	2	0	2	0	0	0	0	0	0	1	1	1
e_{15}	0	0	1	1	0	1	1	0	0	0	0	-1	2	2	0	0	0	0	0	0	0	1	1	1
e_{16}	1	1	0	0	1	0	-1	0	0	0	0	1	0	0	0	0	2	2	1	1	1	0	0	0
e_{17}	1	1	0	0	1	0	-1	0	0	0	0	1	0	0	0	2	0	2	1	1	1	0	0	0
e_{18}	1	1	0	0	1	0	-1	0	0	0	0	1	0	0	0	2	2	0	1	1	1	0	0	0
e_{19}	0	0	0	1	0	-1	0	1	1	0	0	1	0	0	0	1	1	1	0	2	2	0	0	0
e_{20}	0	0	0	1	0	-1	0	1	1	0	0	1	0	0	0	1	1	1	2	0	2	0	0	0
e_{21}	0	0	0	1	0	-1	0	1	1	0	0	1	0	0	0	1	1	1	2	2	0	0	0	0
e_{22}	0	0	0	0	-1	1	1	0	0	1	1	0	1	1	1	0	0	0	0	0	0	0	2	2
e_{23}	0	0	0	0	-1	1	1	0	0	1	1	0	1	1	1	0	0	0	0	0	0	2	0	2
e_{24}	0	0	0	0	-1	1	1	0	0	1	1	0	1	1	1	0	0	0	0	0	0	2	2	0

Let $\pi = \{E_1, E_2, E_3, E_4\}$ be the natural partition for D with the directed edges associated with s_k placed in E_k. Notice that the only entries of 2 occur near the diagonal in

$E_i \times E_i$ blocks. Regardless of how the six edges are enumerated within each partition class, the edge partition π has a quotient matrix:

$$A_E(D/\pi) = \begin{array}{c} \\ E_1 \\ E_2 \\ E_3 \\ E_4 \end{array} \begin{array}{cccc} E_1 & E_2 & E_3 & E_4 \\ \left[\begin{array}{cccc} 0 & -1 & 3 & 0 \\ -1 & 0 & 0 & 3 \\ 3 & 0 & 4 & 3 \\ 0 & 3 & 3 & 4 \end{array}\right] \end{array}$$

Since $A_E(D/\pi)$ is symmetric, Theorem 6.2.2 guarantees that this is a Kirchhoff partition, and therefore G must be a Kirchhoff graph, which of course was already established.

6.3 Equitable and Kirchhoff are not equivalent

Theorems 6.2.1 and 6.2.2 prove that any equitable edge partition with symmetric quotient matrix $A_E(D/\pi)$ is a Kirchhoff partition, and if D is the associated multidigraph for some cyclic vector graph G and π is the natural edge partition, then G is a Kirchhoff graph. The converse, however, is *not* true: *not* every Kirchhoff graph corresponds to an equitable edge partition. Any single Kirchhoff cycle with multiple edge vectors is likely not equitable. In addition, the condition that $A_E(D/\pi)$ be symmetric is necessary; equitability by itself is not sufficient. The next two examples show these results: Example 6.3 shows that not every Kirchhoff graph or Kirchhoff edge partition is equitable, while conversely, Example 6.5 shows that not every equitable edge partition is Kirchhoff.

Example 6.3. Consider the multidigraph shown in Figure 6.5 and the natural edge partition $\pi = (E_1, E_2, E_3)$, where:

$$E_1 = \{e_1, e_2\}, \quad E_2 = \{e_3, e_4\}, \quad E_3 = \{e_5, e_6\} \tag{6.16}$$

Let T be the characteristic matrix of partition π; for all vertices v:

$$\lambda(v)T \in \{[1 \quad 0 \quad 2], [0 \quad 0 \quad 0], [-1 \quad 2 \quad 0], [0 \quad -2 \quad -2]\}$$

and for all cycles C:

$$\chi(C)T \in \{[0 \quad 0 \quad 0], [2 \quad 1 \quad -1]\}$$

Therefore the edge partition π is Kirchhoff. This partition, however, is *not* equitable: edges e_1 and e_2, for example, belong to partition cell E_1, but e_1 is incident to two edges of cell E_3 while edge e_2 is incident to none.

Of course, the Kirchhoff graph corresponding to the multidigraph in Figure 6.5 is also not self-chiral along with the multidigraph not having an equitable natural edge

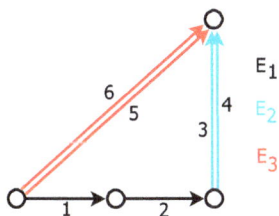

Figure 6.5: An associated multidigraph for a Kirchhoff graph whose natural edge partition is not equitable. Here $m^* = 2$. This edge partition is of course a Kirchhoff edge partition.

partition. If a Kirchhoff graph is symmetric in the sense of being self-chiral, the associated multidigraph might be expect to be equitable natural edge partition. This is indeed the case for the Kirchhoff graph in Example 6.1. In general, however, such a result does not hold, as the next example shows.

Example 6.4. Consider the row matrix

$$R = \begin{bmatrix} 1 & 0 & 1 & 3 \\ 0 & 1 & 1 & 1 \end{bmatrix}$$

and a Kirchhoff graph G for R in Figure 6.6. Let D be the associated multidigraph for G; the natural edge partition for D and for this Kirchhoff graph is not equitable: Notice that all the entries in the first row of $A_E(D)$ corresponding to the fourth partition class are 1, whereas all the entries in the third row of $A_E(D)$ corresponding to the fourth partition class are 0. Thus the definition of the quotient matrix $A_E(D/\pi)$ is impossible, and this natural edge partition is not equitable.

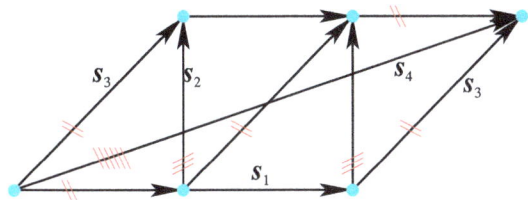

Figure 6.6: A Kirchhoff graph that is self-chiral, but whose associated multidigraph has a natural edge partition that is not equitable. This is a strut Kirchhoff graph with six copies of the fourth edge vector existing the vertex on the lower left and entering the vertex on the upper right

Example 6.5. Now consider the digraph D in Figure 6.7 and edge partition $\pi = \{E_1, E_2, E_3, E_4\}$ with:

$$E_1 = \{e_1, e_2\}, \quad E_2 = \{e_3, e_4\}, \quad E_3 = \{e_5\}, \quad E_4 = \{e_6\}$$

Let T be the characteristic matrix of π, and let C be the cycle v_1, v_2, v_4, v_1. Consider the (non)orthogonality condition:

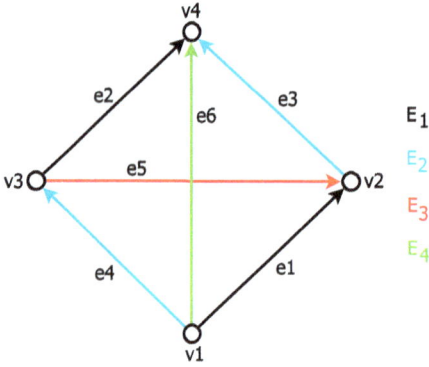

Figure 6.7: A digraph with an edge partition that is equitable, but not Kirchhoff. Edge classes E_1 and E_2 each have two edges, but E_3 and E_4 each have only one.

$$\lambda(v_1)T \cdot \chi(C)T = \begin{bmatrix} 1 & 1 & 0 & 1 \end{bmatrix} \cdot \begin{bmatrix} 1 & 1 & 0 & -1 \end{bmatrix} = 1 \neq 0$$

Thus this edge partition π is not Kirchhoff even though π *is* equitable with a quotient matrix:

$$A_E(D/\pi) = \begin{array}{c} \\ E_1 \\ E_2 \\ E_3 \\ E_4 \end{array} \begin{array}{c} \begin{matrix} E_1 & E_2 & E_3 & E_4 \end{matrix} \\ \begin{bmatrix} 0 & 0 & 1 & 1 \\ 0 & 0 & -1 & 1 \\ 2 & -2 & 0 & 0 \\ 2 & 2 & 0 & 0 \end{bmatrix} \end{array} \qquad (6.17)$$

that of course is *not* symmetric.

The preceding examples demonstrate a lack of exact correspondence between equitable and Kirchhoff edge partitions. A natural next question is to ask if a partial converse to Theorem 6.2.2 holds, that is, if an edge partition π is *both* equitable and Kirchhoff, then is the matrix $A_E(D/\pi)$ symmetric? This situation is addressed next.

6.4 Uniform edge partitions

Theorem 6.2.2 showed that if D is a digraph with equitable edge partition π and the quotient matrix $A_E(D/\pi)$ is symmetric, then π is Kirchhoff. The converse to Theorem 6.2.2, however, is false, as illustrated by Example 6.6.

Example 6.6. Consider the digraph D, with edge partition π presented in Figure 6.8. It has nine directed edges, partitioned into six cells $E_1, \ldots, E_6$. For any vertex v and cycle C, $\lambda(v)T \in \text{Row}(B)$ and $\chi(C)T \in \text{Row}(M^T)$, where B and M^T are the following matrices:

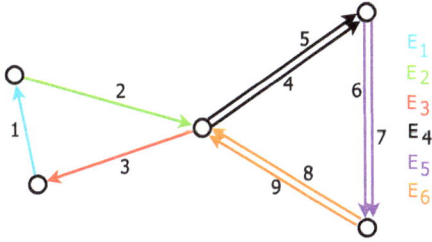

Figure 6.8: An associated multidigraph whose equitable edge partition that is also a Kirchhoff partition, but whose quotient matrix is *not* symmetric. Notice that the vector graph corresponding to this digraph is not cyclic.

$$
B = \begin{array}{c}
\begin{array}{cccccc} E_1 & E_2 & E_3 & E_4 & E_5 & E_6 \end{array} \\
\left[\begin{array}{cccccc}
-1 & 1 & 0 & 0 & 0 & 0 \\
1 & 0 & -1 & 0 & 0 & 0 \\
0 & -1 & 1 & 2 & 0 & -2 \\
0 & 0 & 0 & -2 & 2 & 0 \\
0 & 0 & 0 & 0 & -2 & 2
\end{array}\right]
\end{array}, \quad
M^T = \begin{array}{c}
\begin{array}{cccccc} E_1 & E_2 & E_3 & E_4 & E_5 & E_6 \end{array} \\
\left[\begin{array}{cccccc}
1 & 1 & 1 & 0 & 0 & 0 \\
0 & 0 & 0 & 1 & 1 & 1
\end{array}\right]
\end{array}
$$

As the product BM is the 5×2 zero matrix, the definition of a Kirchhoff partition (6.10) is satisfied for all vertices and cycles. Therefore the edge partition π is Kirchhoff. Moreover, π is equitable and has the quotient matrix

$$
A_E(D/\pi) = \begin{array}{c}
\begin{array}{cccccc} E_1 & E_2 & E_3 & E_4 & E_5 & E_6 \end{array} \\
\begin{array}{c} E_1 \\ E_2 \\ E_3 \\ E_4 \\ E_5 \\ E_6 \end{array}
\left[\begin{array}{cccccc}
0 & -1 & -1 & 0 & 0 & 0 \\
-1 & 0 & -1 & -2 & 0 & 2 \\
-1 & -1 & 0 & 2 & 0 & -2 \\
0 & -1 & 1 & 2 & -2 & -2 \\
0 & 0 & 0 & -2 & 2 & -2 \\
0 & 1 & -1 & -2 & -2 & 2
\end{array}\right]
\end{array}
$$

that is not symmetric.

Although Example 6.6 shows that the converse of Theorem 6.2.2 does not hold in general, a partial converse (Corollary 6.4.2) can be obtained by considering the sizes of partition cells. Let π be any edge partition with k partition cells and characteristic matrix T. Let Λ be a $k \times k$ diagonal matrix with $\Lambda_{i,i} = 1/|E_i|$. Since $T^T T$ is a $k \times k$ diagonal matrix with (i, i)-entry $|E_i|$,

$$
\Lambda T^T T = I_k \tag{6.18}
$$

that is, the matrix ΛT^T is a left pseudoinverse of T. Moreover, if π is equitable, then by (6.6):

$$A_E(D/\pi) = \Lambda T^{\mathsf{T}} T A_E(D/\pi) = \Lambda T^{\mathsf{T}} A_E(D) T \tag{6.19}$$

There is a special distinction for those edge partitions with $|E_i|$ independent of i:

Definition. An edge partition is **uniform** if all partition cells are the same size.

Theorem 6.4.1. *Let D be a connected digraph with equitable edge partition π. Then $A_E(D/\pi)$ is symmetric if and only if π is uniform.*

Proof. First suppose π is uniform. Then $\Lambda = cI$ for some constant c, and

$$A_E(D/\pi) = cT^{\mathsf{T}} A_E(D) T$$

Therefore

$$\begin{aligned}
A_E(D/\pi)^{\mathsf{T}} &= \left(cT^{\mathsf{T}} A_E(D) T\right)^{\mathsf{T}} \\
&= cT^{\mathsf{T}} A_E(D)^{\mathsf{T}} T \\
&= cT^{\mathsf{T}} A_E(D) T = A_E(D/\pi)
\end{aligned}$$

and $A_E(D/\pi)$ is symmetric.

Conversely, suppose that $A_E(D/\pi)$ is symmetric. Then by (6.19), since $A_E(D)$ and Λ are both symmetric,

$$\Lambda\left(T^{\mathsf{T}} A_E(D) T\right) = A_E(D/\pi) = A_E(D/\pi)^{\mathsf{T}} = \left(\Lambda T^{\mathsf{T}} A_E(D) T\right)^{\mathsf{T}} = \left(T^{\mathsf{T}} A_E(D) T\right)\Lambda$$

Therefore for all $i \neq j$,

$$\Lambda_{i,i}\left(T^{\mathsf{T}} A_E(D) T\right)_{i,j} = \left(\Lambda T^{\mathsf{T}} A_E(D) T\right)_{i,j} = \left(T^{\mathsf{T}} A_E(D) T\Lambda\right)_{i,j} = \left(T^{\mathsf{T}} A_E(D) T\right)_{i,j}\Lambda_{j,j} \tag{6.20}$$

Because D is connected, no simultaneous permutation of rows and columns can transform $T^{\mathsf{T}} A_E(D) T$ into a block-diagonal form:

$$\begin{bmatrix} A & 0 \\ 0 & B \end{bmatrix}$$

Therefore as (6.20) is true for all $i \neq j$, it follows that all diagonal entries of Λ must be equal. Therefore $|E_i|$ is independent of i, and π is uniform. □

Let D be a connected digraph with equitable edge partition π. Corollary 6.4.1 is an immediate consequence of Theorem 6.4.1, and Corollary 6.4.2 is the desired partial converse of Theorem 6.2.2.

Corollary 6.4.1. *Every uniform equitable edge partition of D is Kirchhoff.*

Corollary 6.4.2. *If an equitable edge partition π is Kirchhoff and uniform, then the quotient matrix $A_E(D/\pi)$ is symmetric.*

7 Kirchhoff graph duality and Maxwell reciprocal diagrams

Earlier, Kirchhoff graph duality was introduced in terms of exchanging the roles of the row and null spaces for a given matrix. Here we explore this duality in more detail. It is the key thing in understanding how Kirchhoff graphs are related to Maxwell reciprocal diagrams. These diagrams were defined by James Clerk Maxwell [18] in 1870 as a type of geometric reciprocity having applications in mechanics.

This chapter first considers Kirchhoff graphs and their duals, and how these Kirchhoff duals relate to the standard definition of graph duality for planar graphs. It then discusses Maxwell reciprocal diagrams, explaining how to construct them and exploring their connection to Kirchhoff graphs. At first the simplest cases are considered, but later, more complicated cases are dealt with. The final section considers the Kirchhoff duals of Kirchhoff graphs whose associated digraphs have no geometric dual, but where the dual graphs themselves do.

This is in no way a thorough or complete discussion of Maxwell reciprocal diagrams; rather it is just a basic description, sufficient to understand them relative to Kirchhoff graphs. In particular, there is no description here of their implications for mechanics. A somewhat more extensive discussion of these diagrams is given by Reese et al. [24] (2016). Beyond this, there is of course also the original work of Maxwell [18].

7.1 Kirchhoff graph duals and planar duals

This section first discusses Kirchhoff duality and then connects this duality to the duality of planar vector graphs.

7.1.1 Kirchhoff duals

As was mentioned above, Kirchhoff graph duality is based on exchanging the roles of the row and null spaces of a given matrix A. This sort of exchange is perhaps best seen in terms of two examples:

Example 7.1. Consider first the Kirchhoff graph in Figure 7.1(a). As was discussed in Example 1.1, there are canonical row and null matrices for this Kirchhoff graph:

$$R = \begin{bmatrix} 2 & 0 & 1 & 1 \\ 0 & 2 & 1 & -1 \end{bmatrix} \quad \text{and} \quad N = \begin{bmatrix} 1 & 1 \\ 1 & -1 \\ -2 & 0 \\ 0 & -2 \end{bmatrix} \tag{7.1}$$

https://doi.org/10.1515/9783111408576-007

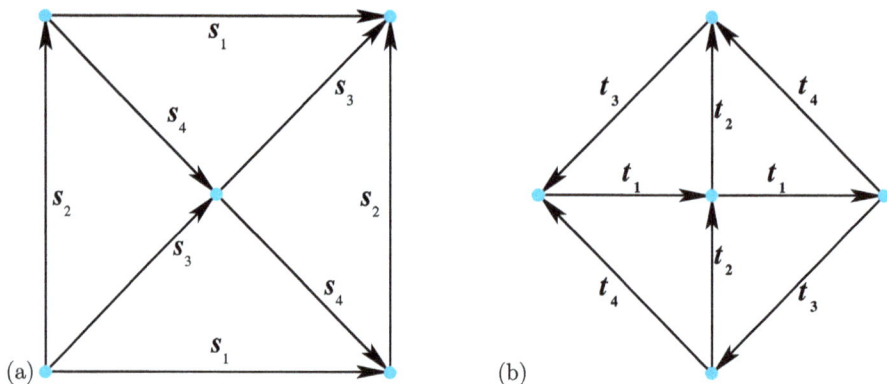

Figure 7.1: Kirchhoff duals. Beginning with (a) the Kirchhoff graph on the left, we can find its cut space and the cycle space, then exchange these spaces to find the cut space and the cycle space of (b) the Kirchhoff graph on the right. The edge vectors in the dual are numbered consistently with the edge vectors in the original Kirchhoff graph. Both Kirchhoff graphs are symmetric and planar with multiplicity $m = 2$.

Now reverse the roles of the row and null spaces: Define $A' := N^{\mathrm{T}}$. Through elementary row operations, A' can be used to produce the dual row and null matrices R' and N' in canonical form:

$$R' = \begin{bmatrix} 1 & 0 & -1 & -1 \\ 0 & 1 & -1 & 1 \end{bmatrix}, \quad N' = \begin{bmatrix} -1 & -1 \\ -1 & 1 \\ -1 & 0 \\ 0 & -1 \end{bmatrix} \tag{7.2}$$

These matrices can be used to construct the dual Kirchhoff graph G' in Figure 7.1(b). The edge vectors in this dual are t_i, $1 \le i \le 4$. Notice that both these Kirchhoff graphs are symmetric, planar, and have multiplicity $m = 2$. As will be made clear later, these Kirchhoff duals are also dual in the standard sense for planar graphs. Keep in mind, however, that if the alternate Kirchhoff graph (the Diamond) was chosen for (7.1), it would have the same Kirchhoff dual even though it is not planar.

Example 7.2. Consider another matrix:

$$A = \begin{bmatrix} 2 & 1 & 0 & 2 & 1 \\ 2 & 2 & 1 & 3 & 1 \\ 0 & 1 & 1 & 0 & 1 \end{bmatrix}$$

A is row-equivalent to the row matrix R, which in turn can be used to derive the null matrix N in canonical form:

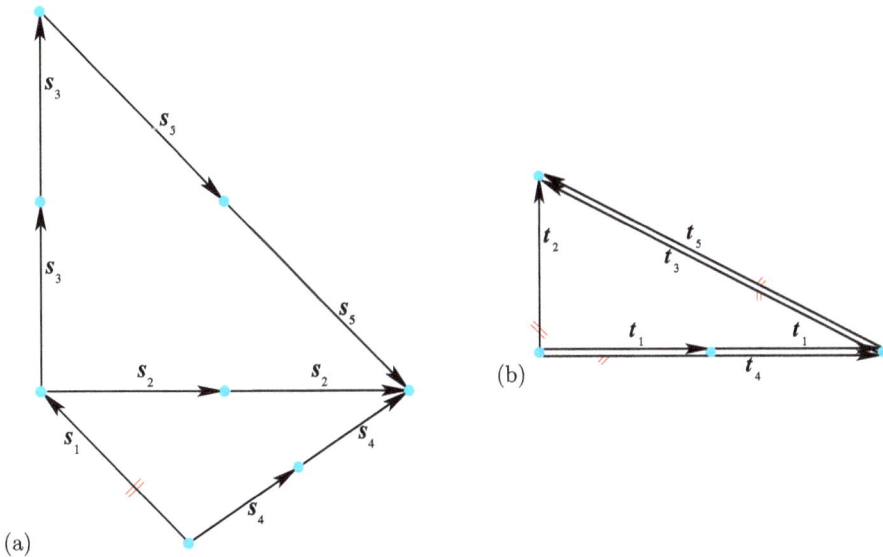

Figure 7.2: Two additional Kirchhoff duals. Beginning with (a) the Kirchhoff graph on the left, one can find its cut and cycle spaces, then exchange these spaces to find the cut and cycle spaces of (b) the Kirchhoff graph on the right. Edge vectors are again labeled consistently between the Kirchhoff graph and its dual. These Kirchhoff graphs are planar but not symmetric. Again, the multiplicity is $m = 2$.

$$
R = \begin{bmatrix} 2 & 0 & 0 & 1 & 0 \\ 0 & 2 & 0 & 2 & 2 \\ 0 & 0 & 2 & 0 & -2 \end{bmatrix}, \quad N = \begin{bmatrix} 1 & 0 \\ 2 & 2 \\ 0 & -2 \\ -2 & 0 \\ 0 & -2 \end{bmatrix}
$$

From these matrices one can construct the Kirchhoff graph G shown in Figure 7.2(a). Notice that based on R, since $k = 3$, this Kirchhoff graph naturally embeds in $\mathbb{R}^3$, but still it can be drawn nicely in a plane.

Now again reverse the roles of the row and null spaces: Again let $A' := N^{\mathsf{T}}$. Again through elementary row operations, A' can be used to produce the dual row and null matrices R' and N' in canonical form:

$$
R' = \begin{bmatrix} 1 & 0 & 2 & -2 & 2 \\ 0 & 1 & -1 & 0 & -1 \end{bmatrix}, \quad N' = \begin{bmatrix} 2 & -2 & 2 \\ -1 & 0 & -1 \\ -1 & 0 & 0 \\ 0 & -1 & 0 \\ 0 & 0 & -1 \end{bmatrix}
$$

Again these matrices can be used to construct the dual Kirchhoff graph G' in Figure 7.2(b). Notice that these Kirchhoff graphs in Figure 7.2 are planar; again the multiplicity for each is $m = 2$. These Kirchhoff duals are again also dual to each other in the standard sense for planar graphs. Notice that having two copies of an edge vector in parallel connecting the same two vertices in either a Kirchhoff graph or its Kirchhoff dual implies that the corresponding edge vector appears in series creating a vertex of degree two in the other graph. Also notice that $t_3 = t_5$ and $t_4 = -2t_1$ in G' even though this is certainly not true in G.

One immediate observation from the previous examples is that the two Kirchhoff graphs basically seem to satisfy the standard classical definition of graph duality for planar graphs and digraphs, though with the dual rotated. So this example leads to the following definitions for vector graphs.

Definition. Two Kirchhoff graphs are **dual** to each other if the cut space of each is the cycle space of the other.

Remark. The definition of *dual* does not imply that the edge vectors themselves are the same in a Kirchhoff graph and its dual; indeed, this is clear in Example 7.1 and Example 7.2. Rather it says that if the edge vectors in the original Kirchhoff graph are labeled $s_1, s_2, \ldots, s_n$ and those in the Kirchhoff dual are labeled consistently $t_1, t_2, \ldots, t_n$, then the cut and cycle spaces are exchanged between the two graphs. In Example 7.2, the cycle space of G and the cut space of G' are the same two-dimensional subspace of $\mathbb{Z}^5$, whereas the cut space of G and the cycle space of G' are the same three-dimensional subspace of $\mathbb{Z}^5$. In the definition for a planar dual discussed below, the term *consistently* means that if the edge vector s_j is bordered by a certain two faces in the original planar Kirchhoff graph, then the edge t_j connects the two vertices in the Kirchhoff dual associated with these faces, implying that the two Kirchhoff graphs have the same number of edge vectors.

Definition. A vector graph is **planar** if it can be embedded in a plane, that is, it can be drawn respecting the vector nature of the edges so that the edge vectors intersect only at vertices.

Remark. The concept of planarity already appeared in the Introduction in Example 1.1. There the Square in Figure 1.1 is planar, whereas the Diamond in Figure 1.2 is not. Indeed, the Diamond is only nonplanar as a vector graph; if it is viewed as a digraph, then the bottom vertex can be moved above the s_0 edge vector, leading to a plane graph. The difference, of course, is that this move is inconsistent with the fact that the edges are vectors.

7.1.2 Duals for planar vector graphs

What is interesting about planar vector graphs is, of course, that there is a standard definition of *dual* for planar digraphs that can be extended to planar vector graphs, and if such a vector graph is a Kirchhoff graph, then this standard definition matches the one above for Kirchhoff graphs.

Definition. For a planar digraph, a (standard) **dual digraph** is obtained from the original digraph by assigning a vertex to each face, including the outer face, and then adding a *consistently* directed edge between vertices if the corresponding faces in the original digraph are adjacent.

The next example demonstrates the equivalence of these two definitions for *dual* (Kirchhoff dual and planar dual) and illustrates a method for constructing a vector graph dual for any planar cyclic vector graph; if this cyclic graph is also a Kirchhoff graph, then the dual constructed here is also a Kirchhoff dual.

Example 7.3. Consider again the Square Kirchhoff graph **G** first discussed in Example 1.1, then further discussed in Example 7.1 and shown in Figure 7.1(a). Its planar dual **G'** can be constructed as follows:

First, construct *D*, the associated planar digraph for the planar vector graph **G** as shown on the left in Figure 7.3, and then form the standard dual digraph by placing a new vertex on each face of the digraph (including the outer face) and adding an edge between vertices on adjacent faces. The directions for these new edges are determined by the directions of the edges in the original digraph: assign directions to edges in the new digraph by traversing cycles *outward* relative to *clockwise* orientation for the original digraph and *inward* relative to *counterclockwise* orientation. This construction is shown on the right in Figure 7.3: the new dual digraph is in red, and as currently drawn, this dual digraph is clearly not a vector graph. The edge labels in both digraphs match the edge vector labels in the corresponding vector graphs; there is no need to introduce separate, unique digraph edge labels in this construction.

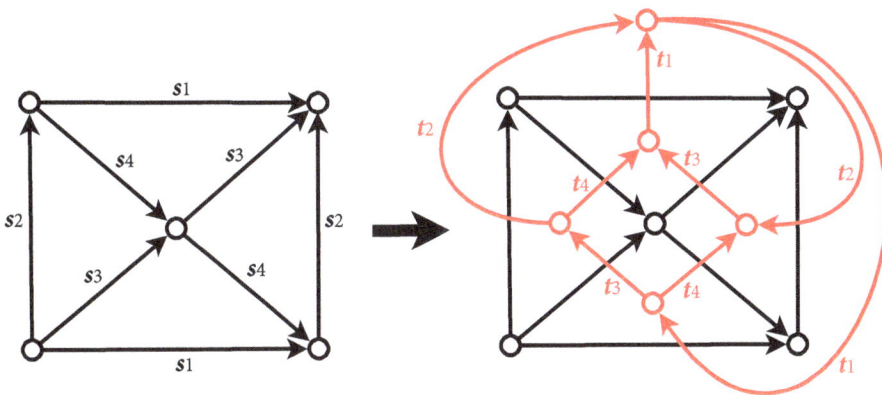

Figure 7.3: Planar digraph dual. Here *D* on the left is the associated digraph with edges labeled by the vector labels from **G** in Figure 7.1(a). On the right, the planar dual *D'* is shown in red, with edge labels derived from the original edge vector labels.

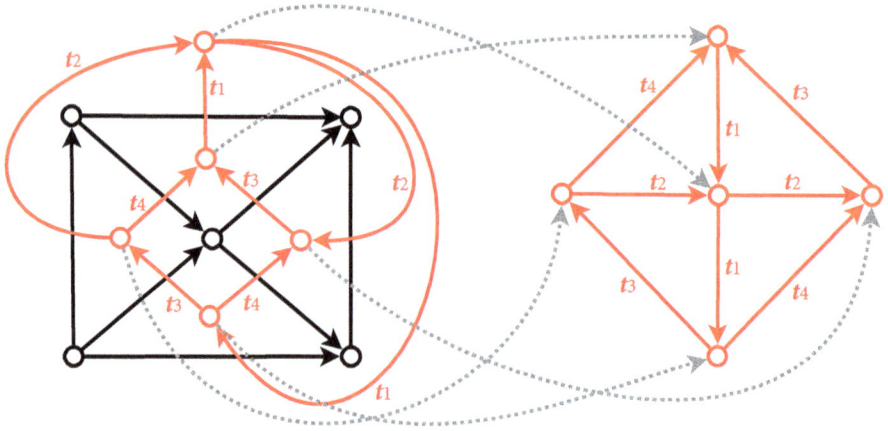

Figure 7.4: Moving from the digraph dual for the associated multidigraph to a version that can be seen directly as the associated multidigraph of G' in Example 7.1.

Although our new dual digraph perhaps does not look like a vector graph, it is possible to recover from it a dual digraph that is immediately equivalent to the Kirchhoff dual G' shown in Figure 7.1(b). To see this, notice that from Example 7.1 the row and null matrices for the Kirchhoff dual are given in (7.2). Now since the columns of R' can be used to represent the vectors in G', the vertex assigned to the outer face can be moved to the center of the graph, retaining the orientation of the directed edges relative to the vertices. The result is a digraph with edges that can be interpreted as vectors, and if this digraph (see Figure 7.4) is rotated 90° counterclockwise, then these vectors match the columns of R'. Hence the resulting vector graph G' is both a standard planar dual and a Kirchhoff dual of G.

The next example deals with duality for asymmetric vector graphs where multiple copies of the same edge vector are incident on the same two vertices or multiple copies appear in series.

Example 7.4. Consider another pair of row and null matrices:

$$R = \begin{bmatrix} 2 & 0 & -1 \\ 0 & 2 & 1 \end{bmatrix}, \quad N = \begin{bmatrix} -1 \\ 1 \\ -2 \end{bmatrix}$$

The multidigraph associated with a single-cycle Kirchhoff graph for this R, N-pair is shown in black on the left in Figure 7.5. To construct the dual multidigraph in this case, a vertex must be assigned to each face, *including* each face between parallel edges incident on the same vertices, and then these vertices must be connected by directed edges when the original faces are adjacent. Again, the direction of the directed edges in the dual multidigraph is outward relative to traversing a cycle clockwise. This initial drawing of the dual multidigraph is shown in blue on the left in Figure 7.5. Now because N has a single column, R' must have a single row, and thus the dual vector graph G' must be one-dimensional. The corresponding one-dimensional dual multidigraph D' is shown on the right in Figure 7.5. Again, the resulting dual vector graph G' is both a planar dual and a Kirchhoff dual.

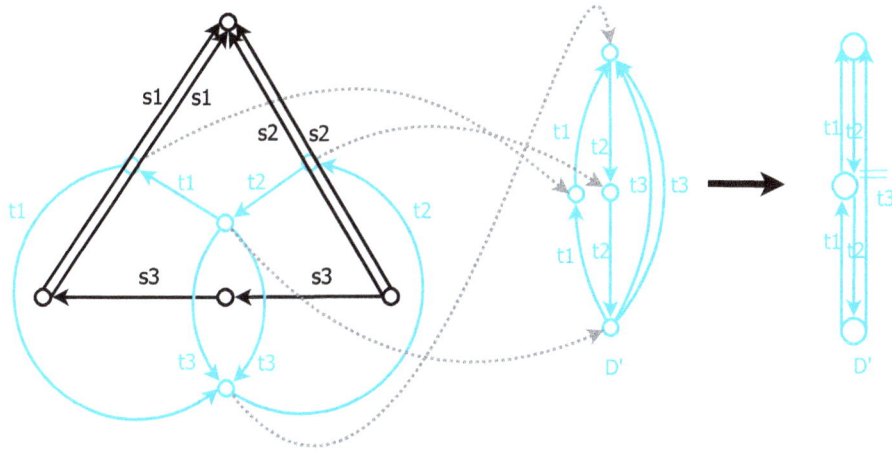

Figure 7.5: Constructing a dual multidigraph for a multidigraph derived from a triangular Kirchhoff graph G having multiple copies of certain edge vectors incident on the same two vertices. The resulting dual Kirchhoff graph G' is degenerated with all edge being scalar multiples of each other.

In the previous two examples, the planar dual of two planar Kirchhoff graphs was also the Kirchhoff dual. The obvious question is whether this is always the case. In fact, it is, as the next theorem states.

Theorem 7.1.1. *For any planar Kirchhoff graph, its dual as a planar vector graph is also a Kirchhoff dual.*

Proof. By construction each vertex in the digraph associated with the original planar Kirchhoff graph corresponds to a cycle in the planar dual digraph, whereas each cycle bordering a face in the original planar Kirchhoff graph corresponds to a vertex in the dual digraph. Since the cycles bordering faces span the cycle space, and since any representation of the edge vectors for the planar vector graph using the columns of a matrix that is row equivalent to the dual row matrix R' must add to zero if and only if these edge vectors form a cycle, the resulting dual vector graph must be a Kirchhoff graph. This final point is just the *consistency condition* for vector graphs. □

Remark. Theorem 7.1.1 shows a strong connection between Kirchhoff duals and the traditional duals for planar graphs. It is worth noting again, however, that Kirchhoff duality extends the notion of duality in a significant way. For example, since the Kirchhoff graph in Figure 7.1b is the dual (both Kirchhoff and planar) of the Square in Figure 7.1(a), it is also the dual (Kirchhoff only) of the Diamond in Figure 1.2.

The connection between the duals of planar vector graphs and Kirchhoff duals leads nicely into a discussion of the connection between these duals and Maxwell reciprocal diagrams, the topic of the next section.

7.2 Maxwell reciprocal diagrams

The key to understanding the Maxwell force balance in mechanics is Maxwell recipro-cal diagrams. Each reciprocal diagram consists of two planar figures that are *reciprocal* to each other, each figure being a *geometric* system of points connected by straight line segments. The two figures in a diagram are the *frame figure* and the *force figure*. The term *geometric* means that the slopes and lengths of the line segments are fixed in each figure. The slopes of corresponding segments must be the same in both figures, that is, each segment in the frame (force) figure is parallel to the corresponding segment in the force (frame) figure. The lengths of the segments in the force figure are proportional to the forces acting on the line segments in the frame figure. For a frame figure to be in equilibrium, when any number of segments meet at a point in the frame figure, the cor-responding segments in the force figure must form a closed polygon. In certain cases, Maxwell observed that if segments meet at a point in the force figure, then the corre-sponding segments in the frame figure also form a closed polygon.

Geometrically, they are considered *reciprocal* if either figure can be taken as a frame figure, and the other as the corresponding force figure, the latter representing the sys-tem of forces that keeps the frame figure in equilibrium. According to Maxwell [18], there is a precise definition of the term *reciprocal*:

Definition. Two plane rectilinear figures are **reciprocal** when they have of an equal number of straight line segments, so that the corresponding parallel segments that meet in a point in one figure form a closed polygon in the other.

The next example, the original example presented by Maxwell, illustrates the con-nections between the two figures in the reciprocal diagram and how to construct a force figure from a frame figure.

Example 7.5. The two figures in Figure 7.6 are the original example of reciprocal figures presented by Maxwell [18]. Note that the same letters are used to label corresponding parallel segments in the frame and force figures. Given a frame figure, Maxwell outlines a method for drawing a force figure corresponding to this frame figure where every segment in the force figure represents, in magnitude and direction, the force acting on a segment in the frame figure. Given the frame figure on the left and assuming that it is in equilibrium, the force figure on the right may be constructed through the following steps:

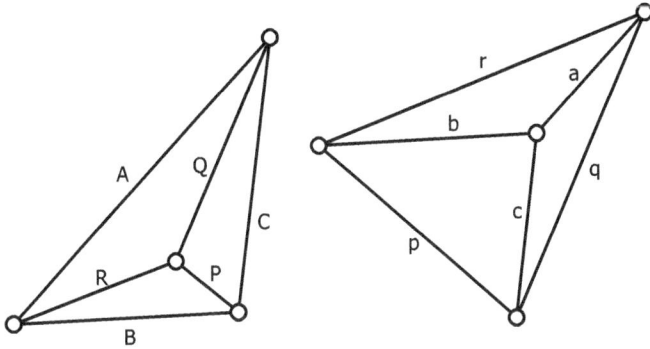

Figure 7.6: A Maxwell reciprocal diagram as presented in [18]. Corresponding segments are labeled with the same letter, uppercase in the frame figure and lowercase in the force figure. Corresponding segments must be parallel.

1. First, draw the segment p in the force figure parallel to the segment P in the frame figure. Choose a length for this segment to represent the magnitude of the force acting along P. The choice of this length is arbitrary, but once it is set, all the other line segment lengths are determined.
2. Forces acting along lines P, Q, and R in the frame figure are in equilibrium. Therefore in the force figure, draw from one endpoint of p a line parallel to Q and from the other endpoint of p a line parallel to R. There are two choices for how to do this; either is acceptable. This construction forms a (closed) triangle pqr in the force figure representing the equilibrium of forces acting on point PQR. Steps 1 and 2 are illustrated in Figure 7.7. Notice that once the length of edge segment p is fixed, the lengths of q and r are determined and represent the magnitude of the forces that must act on Q and R.
3. The other end of segment P in the frame figure meets segments B and C at a point. As the forces at point PBC are also in equilibrium, draw a triangle in the force figure with p as one side and segments parallel to B and C as the others. There are again two ways to construct these segments, illustrated in Figure 7.8, but now the choice matters; only one choice is consistent with this construction leading to a force figure, and one can use the frame figure to determine which of the two choices is the correct one. The endpoints of segment p in the force figure correspond to closed polygons in the frame figure, namely, those polygons that contain P as an edge, polygons PRB and PQC. As PRB is a closed polygon in the frame figure, segment b in the force figure (parallel to B) must meet segments r and p in a point. Similarly, since PQC is a closed polygon in the frame figure, segment c in the force figure (parallel to C) must

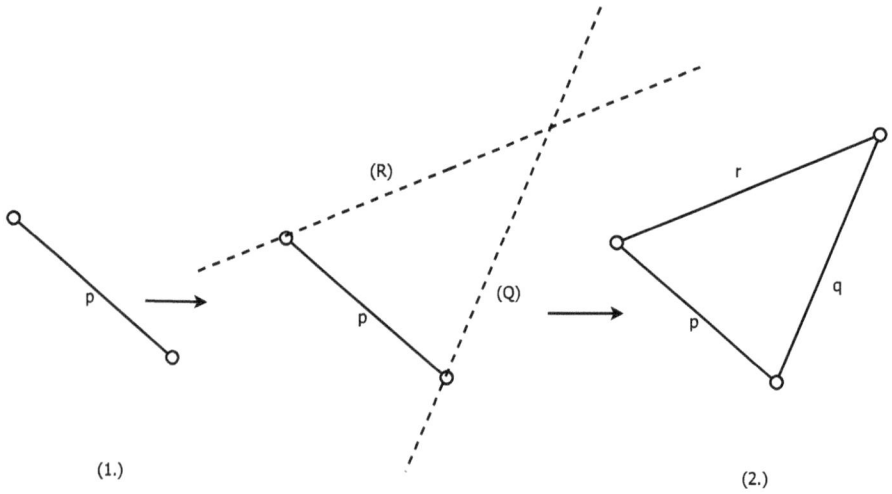

Figure 7.7: Construction Steps 1 and 2 producing the triangle *pqr* in the force figure.

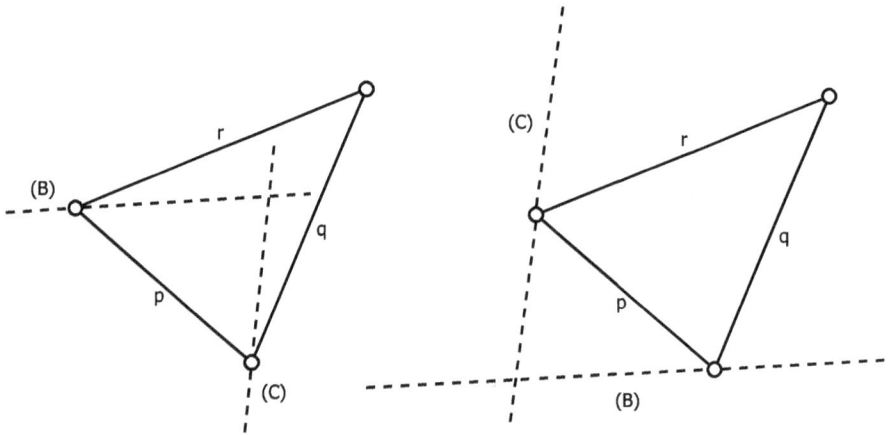

Figure 7.8: Two possible choices for Step 3. The left one is consistent with the existence of triangles *PRB* and *PQC* in the frame diagram; the right one is not.

meet segments *q* and *c* in a point. This corresponds to making the left choice shown in Figure 7.8 and rules out the right choice.

4. Now consider the force equilibrium at point *QCA* in the frame figure. Two of the corresponding segments of force, *q* and *c*, have already been determined in the force figure. Therefore the only choice for segment *a* (parallel to *A*) in the force figure is to complete triangle *qca*. Step 4 is illustrated in Figure 7.9.

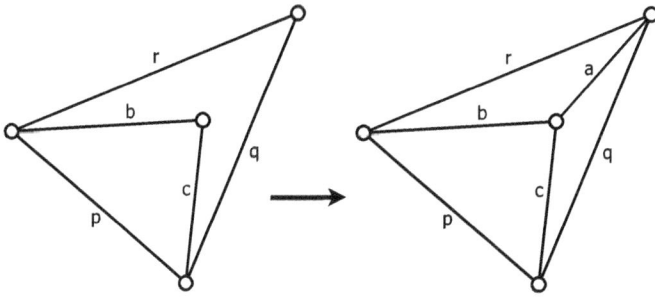

Figure 7.9: Step 4. The addition of line segment *a* completes the force diagram.

Example 7.5 illustrates that the method for constructing a force figure reduces finding the equilibrium of forces acting on the line segments in the frame figure to examining whether the force figure has closed polygons. The figures in Maxwell reciprocal diagrams are in fact geometric graphs that are dual to each other and satisfy the condition that corresponding edges are parallel.

7.2.1 Maxwell figures as polyhedral projections

Beyond the basic description given above, Maxwell also discussed these reciprocal figures in terms of plane projections of (closed) polyhedra. For our purposes, a closed polyhedron is a region bounded by some (finite) number of intersecting planes. What follows here is a step-by-step construction of a Maxwell reciprocal diagram as a polyhedral projection as presented by Maxwell [18].

1. Suppose $\mathcal{P}$ is a closed polyhedron bounded by a set of planes $M_1, \ldots, M_n$, and let $N_1, \ldots, N_n$ be the normal vectors to these planes. Choose a projection plane M that satisfies two conditions:
   (a) M does not intersect the closed polyhedron $\mathcal{P}$.
   (b) For all i, $1 \leq i \leq n$, a line through normal vector N_i of plane M_i intersects plane M.
   The standard plane projection of $\mathcal{P}$ onto M gives one of the two Maxwell figures.

Now construct a *second* polyhedron $\mathcal{P}'$ that with respect to some paraboloid of revolution is a geometric reciprocal to the first polyhedron. The projection of $\mathcal{P}'$ onto projection plane M is then a reciprocal figure of the projection of $\mathcal{P}$ onto M. These two figures are reciprocal in the sense that the corresponding lines are *perpendicular* to each other. We can obtain reciprocal figures with parallel orientation simply by rotating one figure by 90 degrees in the plane. Construct $\mathcal{P}'$ as follows:

2. Let z_0 be a fixed point in three-space that does not lie in M. Take z_0 to be the *origin* for this construction and take the line perpendicular to the projection plane M passing

through z_0 to be the *axis*, denoted as z. Lastly, let z_M be the point of intersection of line z with plane M.

3. For each $k, 1 \le k \le n$, draw a line normal to plane face M_k of $\mathcal{P}$ and passing through point z_0. This line will intersect the projection plane M at a unique point, call it m_k. Each point m_k may be thought of as the projection of face M_k for the second figure onto plane M.

4. For each $k, 1 \le k \le n$, let z_k be the point of intersection of axis z with face M_k. This intersection always exists since M was chosen not to be parallel to any of the normal vectors N_k (and z is a normal line of M). Let d_k be the three-space vector starting at the point z_M and ending at the point z_k. Thus d_k is the vector between the intersection points of the line z with M and M_k. Finally, define the point

$$p_k = m_k - d_k$$

as the "point corresponding to face M_k of the polyhedron $\mathcal{P}$." Steps 2–4 of finding a point p_k corresponding to a particular face M_k are illustrated in Figure 7.10. We have now determined n points $p_1, \ldots, p_n$ corresponding to the faces $M_1, \ldots, M_n$ of $\mathcal{P}$. These points form the vertices of polyhedron $\mathcal{P}'$.

5. For any two faces M_i and M_j that meet in an edge in polyhedron $\mathcal{P}$, draw a line between the corresponding points p_i and p_j. These new lines give us the edges of

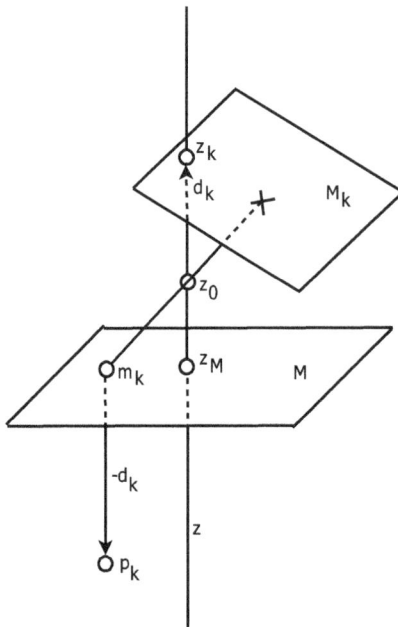

Figure 7.10: The process for finding the point p_k corresponding to a particular face M_k.

polyhedron $\mathcal{P}'$, and the projection of $\mathcal{P}'$ onto M is close to the second Maxwell figure.

6. Now let M_iM_j denote the edge at the intersection of planes M_i and M_j, and let p_ip_j denote the line joining the corresponding points. One can check geometrically that the projections of lines M_iM_j and p_ip_j onto M are perpendicular to each other. The projection of $\mathcal{P}'$ onto M produces a figure where every line is perpendicular to the projection onto M of the corresponding line of $\mathcal{P}$. Moreover, for any lines that meet at a point in one projection, the corresponding lines form a closed polygon in the other projection, that is, the projections onto M of $\mathcal{P}$ and $\mathcal{P}'$ form a Maxwell reciprocal diagram once one of the two figures is rotated by 90 degrees.

7.3 Kirchhoff graphs and Maxwell reciprocal diagrams

Maxwell reciprocal diagrams can be viewed as a pair of dual geometric graphs in which the vertex cuts in each graph correspond to the cycles in the other. This section discusses the relationship between these dual geometric graphs and Kirchhoff graphs, particularly Kirchhoff duals, first for planar Kirchhoff graphs, then more generally.

7.3.1 Planar Kirchhoff graphs and reciprocal diagrams

Suppose that, as in Example 7.5, a Maxwell reciprocal diagram has two figures that, when viewed as graphs, are planar geometric graphs with no edge in either figure parallel to another edge in that figure. This will guarantee that all edges are distinct when viewed as vectors. It also guarantees that all vertices in both Maxwell reciprocal figures have degree three or higher. Now a Kirchhoff graph can be derived from each figure in this Maxwell reciprocal diagram, and these Kirchhoff graphs will be Kirchhoff duals.

To obtain a Kirchhoff graph corresponding to each reciprocal figure, label the edges of the frame figure as e_j and the corresponding parallel edges of the force figure as f_j. Let both vertices and cycles be indexed by these edge labels: $\{e_i, e_j, e_k\}$ denotes the vertex in the frame figure incident to edges e_i, e_j, and e_k, and $f_if_jf_k$ then denotes the corresponding cycle in the force figure (standard cycle notation for simple planar graphs). Recall that $\chi(f_if_jf_k \ldots)$ is the cycle vector for the cycle $f_if_jf_k \ldots$, while $\lambda(\{e_i, e_j, e_k, \ldots\})$ is the vertex cut (incidence vector) for the vertex $\{e_i, e_j, e_k, \ldots\}$. By arbitrarily assigning directions to each edge in the frame figure, construct a planar digraph, E. A planar digraph F corresponding to the force figure can now be constructed by carefully assigning directions to each edge f_i so that:

$$\chi(f_if_jf_k \ldots) = \lambda(\{e_i, e_j, e_k, \ldots\}) \tag{7.3}$$

for each cycle $f_if_jf_k \ldots$ in the force figure, where each component of this common vector is -1, 0, or 1. Finally, define a pair of vector graphs $\mathbf{E}$ and $\mathbf{F}$ from E and F by defining a vector of the same length and direction for each edge in both E and F. Label all the edge vectors so that s_i in E corresponds to e_i in E and t_j in F corresponds to f_j in F. This process results in two vector graphs derived from the original pair of Maxwell figures, and since the edge vectors in each vector graph are distinct, each vector graph is a Kirchhoff graph.

Definition. Let E and F be a pair of **corresponding Kirchhoff graphs** as defined above. Note that there is a nonuniqueness in this definition because of the arbitrary assignment of directions to edges in the frame figure.

Example 7.6. Figure 7.11 displays the Maxwell reciprocal figures as in Figure 7.6, now relabeled as described above. Figure 7.12 then shows a pair of Kirchhoff graphs E and F corresponding to these Maxwell reciprocal figures. Based on the vertex cuts for these two Kirchhoff graphs, the following two matrices can be defined:

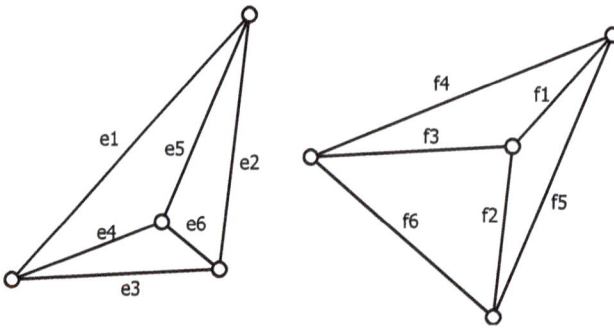

Figure 7.11: Maxwell reciprocal diagram from Figure 7.6 with figures now relabeled as graphs.

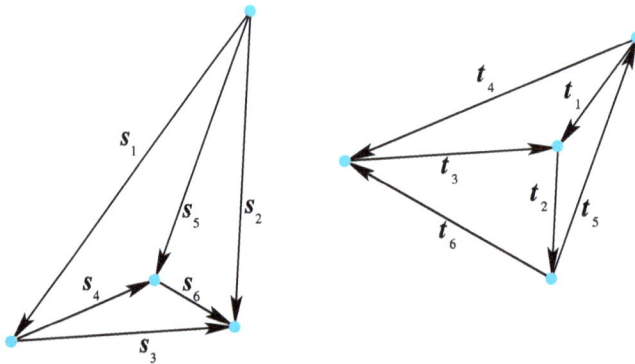

Figure 7.12: Kirchhoff graphs E and F corresponding to Maxwell reciprocal diagram in Figure 7.11. Notice that $s_i \| t_j$.

$$A_E = \begin{bmatrix} 1 & 1 & 0 & 0 & 1 & 0 \\ 0 & 1 & 1 & 0 & 0 & 1 \\ 0 & 0 & 0 & 1 & 1 & -1 \end{bmatrix} \quad \text{and} \quad A_F = \begin{bmatrix} 1 & 0 & 0 & 1 & -1 & 0 \\ 0 & 1 & 0 & 0 & -1 & -1 \\ 0 & 0 & 1 & -1 & 0 & -1 \end{bmatrix}$$

Here each row of each matrix corresponds to one of the four vertex cuts (or its negative) of E or F. In each case, the fourth vertex cut is linearly dependent on the other
three. So E is a Kirchhoff graph for A_E, and F is a Kirchhoff graph for A_F. Moreover,
$\text{Null}(A_E) = (\text{Row}(A_F))^{\text{T}}$, implying that E and F are Kirchhoff duals. Thus beginning with
the figures of a Maxwell reciprocal diagram, and with directions arbitrarily assigned
to the frame figure, assign directions consistently to the force figure and construct two
Kirchhoff graphs, each being the Kirchhoff dual of the other.

The previous example is one case of a more general result:

Theorem 7.3.1. *Suppose that G is a simple, planar Maxwell frame figure viewed as a geo-*
metric graph with no two edges parallel. Also suppose that based on G, one can construct
a corresponding Maxwell force figure H that when viewed as a geometric graph, is also
simple and planar. This force figure H will also have no two edges parallel. Moreover, if E
and F are constructed from G and H as in Example 7.6, then E and F are planar Kirchhoff
graphs and Kirchhoff duals of one another.

Proof. By the definition of a Maxwell reciprocal diagram, if no two edges in G are parallel, then no two edges in H can be parallel either. Also, if no two edges in G are parallel,
then by construction no two vectors s_i in E can be parallel (have the same angle $\theta(s_i)$
with the positive x-axis). The same is true for all the edges t_i in F. Thus all the edge
vectors in each vector graph E and F are distinct, and hence by Theorem 2.3.1 each is a
Kirchhoff graph. Finally, E and F are dual because of the edge assignment

$$e_i \mapsto s_i \quad \text{and} \quad f_i \mapsto t_i \quad \text{for all } i$$

and the vertex-cut-cycle correspondence given in (7.3). In other words, by (7.3), if E corresponds to some matrix A_E, then F corresponds to $\text{Null}(A_E)$. So, indeed, $F = E'$. □

In the previous example, the relabeling of vector edges may seem trivial; in the next
example, this process is definitely not trivial. Indeed, Theorem 7.3.1 above might lead us
to suspect that *any* planar polyhedral projection would produce a Maxwell reciprocal
diagram that leads to a pair of dual Kirchhoff graphs. This is not necessarily the case,
however, because some of the edge vectors in the vector graphs may be identical, even
though no edges in the original polyhedron were parallel. The next example demonstrates this issue.

Example 7.7. In this example a Maxwell frame figure G is constructed as a polyhedral
projection. As before, the reciprocal Maxwell force figure H can be constructed from G.
The vector graph G corresponding to G, however, is *not* a Kirchhoff graph. The vector
graph H corresponding to H, on the other hand, *is* a Kirchhoff graph.

Consider the following system of five planes in $\mathbb{R}^3$:

$$-x + z = 0, \quad -x + 3y - 2z = 0, \quad y + z = 0, \quad 5x - y + 4z = 0, \quad z = 0 \qquad (7.4)$$

The three-dimensional region bounded by these five planes forms a closed bounded polyhedron in $\mathbb{R}^3$, and the projection of this figure onto the (x, y)-plane is a simple, planar graph that can be viewed as a Maxwell frame figure. This polyhedron and its projection onto the (x, y)-plane are shown on Figure 7.13. The polyhedron was very specifically chosen so that its projection onto the (x, y)-plane has two edges that are parallel and have the same length (see Figure 7.13b). As the corresponding edges are not parallel in the polyhedron, it is worth noting that a randomly chosen plane projection of this polyhedron would not result in any edges being parallel with identical lengths.

Viewing the projection in Figure 7.13b as a frame figure, one can now construct the associated force figure. The simple planar graphs G (corresponding to the frame figure)

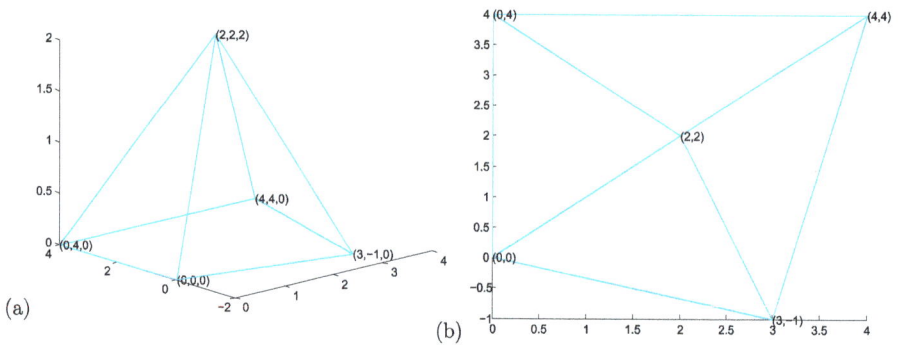

Figure 7.13: (a) A closed polyhedron whose boundary planes are given by the equations in (7.4). (b) The projection of this closed polyhedron onto the (x, y)-plane. Notice that two edges in this projection have identical length and direction.

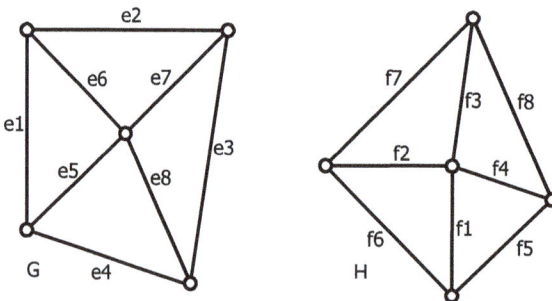

Figure 7.14: Maxwell reciprocal figures G and H corresponding to the closed polyhedron defined by (7.4). Edges e_5 and e_7 in the frame figure are parallel with identical lengths, while edges f_5 and f_7 in the force figure are parallel.

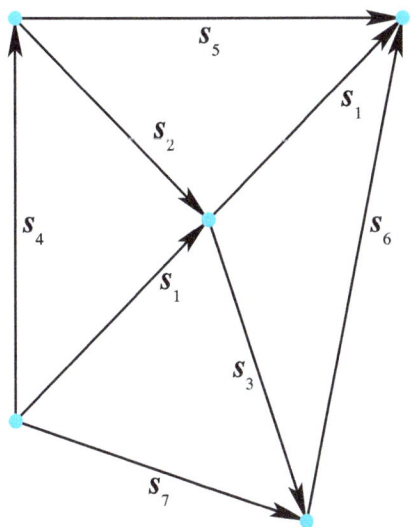

Figure 7.15: One possible vector graph **G** corresponding to the Maxwell reciprocal figures G. For any such vector graph, the same vector would correspond to e_5 and e_7.

and H (corresponding to the force figure) are shown in Figure 7.14. The vector graph **G** shown in Figure 7.15 is one of the possible vector graphs corresponding to the frame diagram G in Figure 7.14. Notice that here edge vectors s_2 through s_7 appear once, but edge vector s_1 appears twice, and because e_5 and e_7 are parallel with the same length, this multiplicity for the corresponding edge vector will always different from all the others. So any vector graph corresponding to G cannot be uniform with respect to edge vector multiplicity and hence cannot be a Kirchhoff graph. On the other hand, consider the force diagram H. For the corresponding vector graph **H** corresponding to H, every edge vector is distinct, so by Theorem 2.3.1, **H** must be a Kirchhoff graph.

Remarks. The previous example illustrates several important points:

1. Not every simple planar vector graph corresponding to a Maxwell figure is a Kirchhoff graph.
2. It is possible that a pair of Maxwell reciprocal figures generates a pair of corresponding simple planar vector graphs, one of which is Kirchhoff, and the other is not. This discrepancy arises because the corresponding vector graphs may have different numbers of distinct vector edges. This difference occurs because although corresponding edges in the frame and force figures must be parallel, their lengths may vary.
3. Example 7.7 does not imply that only simple planar figures with no parallel edges can correspond to a pair of Kirchhoff dual graphs. Symmetry in the frame diagram can lead to symmetry in the force diagram, resulting in a number of identified edges in the corresponding vector graphs. The next example illustrates this point.

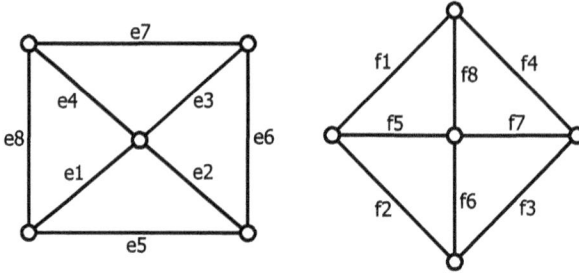

Figure 7.16: A pair of Maxwell reciprocal figures similar to the Kirchhoff duals in Figure 7.1.

Example 7.8. Consider the pair of Maxwell reciprocal figures shown in Figure 7.16 and the corresponding pair of planar vector graphs shown in Figure 7.1. We can verify that the pair in Figure 7.16 forms a Maxwell reciprocal diagram, that each reciprocal figure corresponds to one of the vector graphs, and that these two vector graphs are both Kirchhoff graphs corresponding, respectively, to two matrices:

$$A_e = \begin{bmatrix} 1 & 1 & 1 & 0 \\ -1 & 1 & 0 & 1 \end{bmatrix} \quad \text{and} \quad A_f = \begin{bmatrix} 1 & 0 & -1 & 1 \\ 0 & -1 & 1 & 1 \end{bmatrix}$$

Finally, these Kirchhoff graphs are dual to each other since $\text{Null}(A_e) = (\text{Row}(A_f))^{\text{T}}$.

7.4 Nonplanar vector graphs without geometric duals

Example 7.6 and Example 7.8 illustrate that in many cases, Maxwell reciprocal figures can lead to pairs of Kirchhoff duals. Example 7.7, however, demonstrates that these two theories are not equivalent: vector graphs corresponding to a Maxwell figure are not necessarily Kirchhoff when edges are identified as vectors. Even if the vector graph corresponding to a Maxwell figure *is* Kirchhoff, the vector graph corresponding to the reciprocal figure *need not* be Kirchhoff. Moreover, there are a number of classes of vector graphs that will never correspond to a Maxwell figure; for instance, vector graphs with multiple edges and vector graphs that are not the projection of a polyhedron.

7.4.1 Nonplanar Kirchhoff graphs

The planar Kirchhoff graphs given above are not representative of all Kirchhoff graphs that may be neither be planar nor have a *geometric dual* (a higher dimension version of the planar dual; see Example 7.9 below). Indeed, one advantage of the concept of *Kirchhoff dual* is the opportunity to construct a dual to Kirchhoff graphs in cases where no standard dual vector graph exists. In the following example, a Kirchhoff graph **H** based on the complete bipartite graph $K_{3,3}$ is considered. This Kirchhoff graph has no planar or geometric dual, but since it is Kirchhoff, it has a Kirchhoff dual. From **H** a Kirchhoff dual **H′** and a double dual **H″** are computed. Since the digraph associated with **H′** *is* a geometric digraph, a double dual **H″** can be found through geometric means.

Example 7.9. Consider the vector graph **H** whose associated digraph D is shown in Figure 7.17. The edge vectors in **H** are all distinct, so again by Theorem 2.3.1, **H** must be a Kirchhoff graph, and hence a Kirchhoff dual can be found, even though the associated digraph D (based on the complete bipartite graph $K_{3,3}$) is not planar and has no geometric dual. A digraph D′ associated with **H′**, a Kirchhoff dual of **H**, is shown in Figure 7.18.

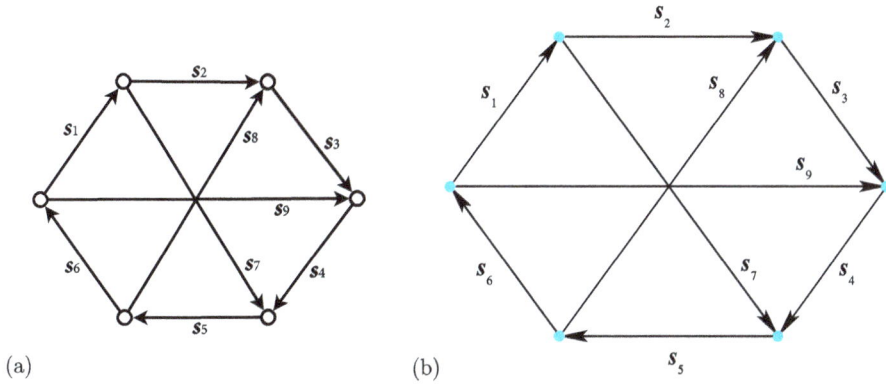

(a)   (b)

Figure 7.17: (a) Digraph D associated with a vector graph **H** based on the bipartite graph $K_{3,3}$. (b) Vector graph **H**. Edge vectors in **H** are distinct; the edge labeling for D reflects the edge vector labels in **H**.

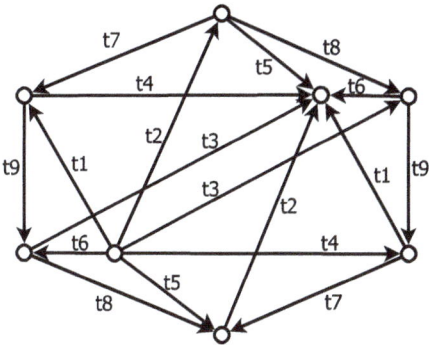

Figure 7.18: Digraph D′ associated with a vector graph **H′**, the Kirchhoff dual of **H**. All edge vectors in **H′** appear twice, and again the edge labelling reflects the labels for the edge vectors in the dual **H′**.

Either D or H can lead to the associated row and null matrices R and N:

$$R = \begin{bmatrix} 1 & 0 & 0 & 0 & 0 & -1 & 0 & 0 & 1 \\ 0 & 1 & 0 & 0 & 0 & -1 & 1 & 0 & 1 \\ 0 & 0 & 1 & 0 & 0 & -1 & 1 & -1 & 1 \\ 0 & 0 & 0 & 1 & 0 & -1 & 1 & -1 & 0 \\ 0 & 0 & 0 & 0 & 1 & -1 & 0 & -1 & 0 \end{bmatrix}, \quad N = \begin{bmatrix} -1 & 0 & 0 & 1 \\ -1 & 1 & 0 & 1 \\ -1 & 1 & -1 & 1 \\ -1 & 1 & -1 & 0 \\ -1 & 0 & -1 & 0 \\ -1 & 0 & 0 & 0 \\ 0 & -1 & 0 & 0 \\ 0 & 0 & -1 & 0 \\ 0 & 0 & 0 & -1 \end{bmatrix}$$

Now since again $A' := N^{\mathrm{T}}$, row reduction yields the matrix R':

$$R' = \begin{bmatrix} 1 & 0 & 0 & 0 & 1 & 1 & 1 & 0 & 0 \\ 0 & 1 & 0 & 0 & -1 & 0 & -1 & -1 & 0 \\ 0 & 0 & 1 & 0 & 0 & -1 & 0 & 1 & -1 \\ 0 & 0 & 0 & 1 & 1 & 1 & 0 & 0 & 1 \end{bmatrix}$$

Since the rows of R' form a basis for the cut space of D' and H', the Kirchhoff graph H' is indeed a Kirchhoff dual of H. Since H' has no Kirchhoff subgraph, it is also prime.

Perhaps the most interesting aspect of this example is that even though H has no geometric dual, H' has: the Kirchhoff dual H' and its associated digraph D' are a plane projection of a polyhedron in $\mathbb{R}^3$, meaning that D' has a geometric dual. This double dual is D''; for each directed edge t_j, $1 \leq j \leq 18$, in D', the corresponding directed edge in D'' is s_j. This polyhedron and its geometric dual are illustrated in Figure 7.19. The directed edges of the polyhedron correspond directly to the edge vectors in H' since identified directed edges have the same length and direction. The geometric double dual D'' on the right in Figure 7.19 is constructed by assigning a vertex to each face of the polyhedron and then connecting vertices with directed edges if the corresponding faces are adjacent. Notice that the digraph D'' does not directly correspond to H'' (or any vector graph) because not all identified edges have the same length or direction.

To construct the double dual vector graph H'', the vertices of D'' must be embedded in the plane so that all directed edges with label s_i have the same length and direction. This can be accomplished by mapping for each j, $1 \leq j \leq 6$, the pair of vertices v_j and v_{j+6} to the same point in the plane. This embedding is illustrated in Figure 7.20. This vector graph is in fact the Kirchhoff graph H, which of course is not a surprise since it is the Kirchhoff double dual of H. So H'' is both a Kirchhoff dual for H' and a double Kirchhoff dual of itself. The differences is just that, as constructed, each edge vector occurs twice, though the proper view would still seem to be that $H'' = H$.

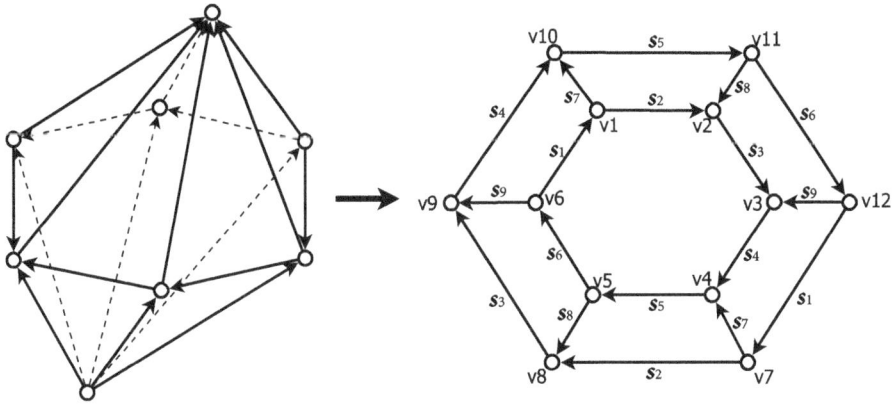

Figure 7.19: Geometric double dual. The digraph on the left is a three-dimensional version of D'. It forms a polyhedron and corresponds directly to H' since identified edges have the same length and direction. On the right is D'', the geometric dual of D', a hence the double dual of D. Each vertex in D'' corresponds to a face of the polyhedron, and directed edges connect these vertices exactly when those faces are adjacent. Again, the edges are labeled with the corresponding edge vector, not the edge enumeration. Notice that as drawn here, D'' does not correspond directly to any vector graph since not all identified edges have the same length or direction.

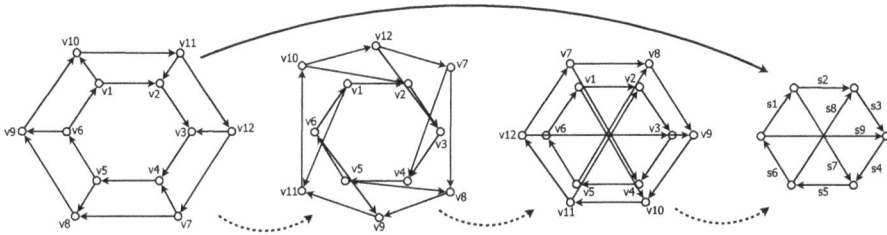

Figure 7.20: Constructing a version of D'' that corresponds directly to H'', rotating from the geometric dual of the polyhedron to two copies of the original digraph D.

By following essentially the reverse process as that outlined in Example 7.9 one can construct a Kirchhoff dual for a Kirchhoff graph based on K_5, the complete graph on five vertices, which of course is also nonplanar.

Example 7.10. Consider the vector graph K whose associated digraph D is shown in Figure 7.21, constructed by assigning edge vectors to K_5. Since all the edge vectors are distinct, K must be a Kirchhoff graph.

As in the previous example, even though K is nonplanar, since all the edge vectors are distinct, the vector graph K must be a Kirchhoff graph. Based on the incidence matrix or the vertex cuts, the corresponding row and null matrices can be derived:

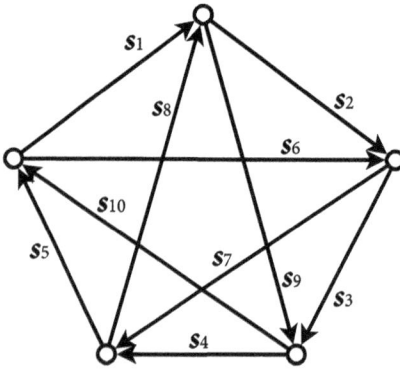

Figure 7.21: An associated digraph D for the Kirchhoff graph K, constructed based on K_5. Again, all edge vectors are distinct, guaranteeing that K is Kirchhoff.

$$R = \begin{bmatrix} 1 & 0 & 0 & 0 & -1 & 1 & 0 & 0 & 0 & -1 \\ 0 & 1 & 0 & 0 & -1 & 1 & 0 & -1 & 1 & -1 \\ 0 & 0 & 1 & 0 & -1 & 0 & 1 & -1 & 1 & -1 \\ 0 & 0 & 0 & 1 & -1 & 0 & 1 & -1 & 0 & 0 \end{bmatrix}$$

$$N = \begin{bmatrix} -1 & 1 & 0 & 0 & 0 & -1 \\ -1 & 1 & 0 & -1 & 1 & -1 \\ -1 & 0 & 1 & -1 & 1 & -1 \\ -1 & 0 & 1 & -1 & 0 & 0 \\ -1 & 0 & 0 & 0 & 0 & 0 \\ 0 & -1 & 0 & 0 & 0 & 0 \\ 0 & 0 & -1 & 0 & 0 & 0 \\ 0 & 0 & 0 & -1 & 0 & 0 \\ 0 & 0 & 0 & 0 & -1 & 0 \\ 0 & 0 & 0 & 0 & 0 & -1 \end{bmatrix}$$

Again, the row matrix R' for a Kirchhoff dual can be obtained from row operations on $A' := N^{\mathsf{T}}$:

$$R' = \begin{bmatrix} 1 & 0 & 0 & 0 & 0 & 0 & 0 & 0 & 1 & 1 \\ 0 & 1 & 0 & 0 & 0 & 0 & 1 & 1 & 0 & 0 \\ 0 & 0 & 1 & 0 & 0 & 0 & -1 & -1 & -1 & 0 \\ 0 & 0 & 0 & 1 & 0 & 0 & 0 & 1 & 1 & 0 \\ 0 & 0 & 0 & 0 & 1 & 0 & 0 & -1 & -1 & -1 \\ 0 & 0 & 0 & 0 & 0 & 1 & 1 & 1 & 1 & 1 \end{bmatrix} \tag{7.5}$$

A Kirchhoff dual could be constructed from R' as was done for previous examples, but since any Kirchhoff graph would have at least six vertices, such a construction might be difficult. An easier approach might be to reverse the approach for finding the dou-

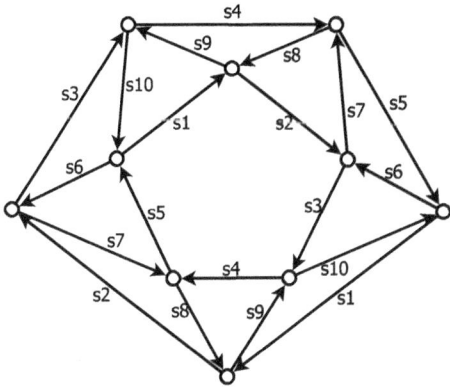

Figure 7.22: Digraph D, an *unfolding* of the associated digraph for the Kirchhoff graph K.

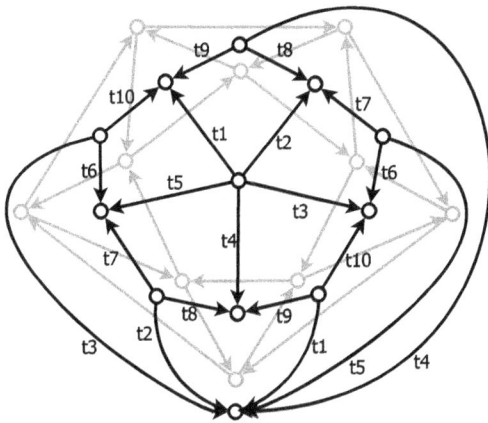

Figure 7.23: Constructing the dual D' from digraph D. Gray edges are the geometric digraph D as drawn in Figure 7.22; black edges are D', drawn connecting the faces of D.

ble dual in Example 7.9: Double the vertices and directed edges for the digraph in Figure 7.21, then move one set of vertices outward, rotate this set $180°$, and preserve the digraph connectivity as in Figure 7.22. Next notice that although it does not directly correspond to a vector graph, the digraph in Figure 7.22 is planar and thus has a planar dual; this planar dual D' is shown in Figure 7.23. Again, as drawn, D' in Figure 7.23 does not directly correspond to a vector graph; however, it can be viewed as a geometric digraph whose edges when embedded in $\mathbb{R}^3$ form a polyhedron similar to that on the left in Figure 7.19. This geometric version of D' is shown in Figure 7.24. Now if the digraph edges are replaced by edge vectors, then the result is the desired Kirchhoff dual K'. Indeed, the vector graph K' is a Kirchhoff graph generated by R' in (7.5), and thus K' has been constructed geometrically from $K = K''$.

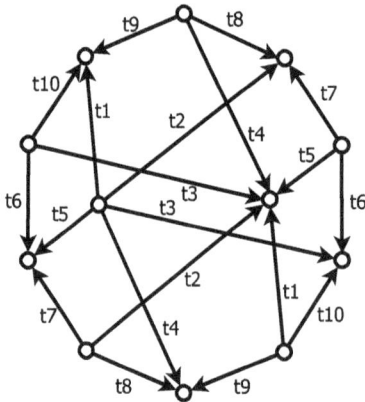

Figure 7.24: The dual D' from Figure 7.23, now drawn as a geometric digraph. From here, the Kirchhoff dual K' can be found by replacing each directed edge by the corresponding edge vector.

7.5 Kirchhoff duals: open questions

This chapter has given a brief introduction to the relationship between Kirchhoff duals, planar and geometric duals, and Maxwell reciprocal diagrams. This relationship has been explored mainly through a set of examples, though the equivalence of Kirchhoff duals on the one hand and planar and geometric duals on the other has been proven for vector graphs and their associated digraphs when both types of duals can be defined. Planar Maxwell reciprocal diagrams were also shown to lead to Kirchhoff duals, at least when no edges are parallel inside either reciprocal figure of the diagram. Nonetheless the examples discussed here suggest that much more work is needed on the relationships amongst these three concepts. For example, starting with a Kirchhoff graph G', is it always possible to *unfold* some version of an associated multidigraph D to find a planar or geometric graph that allows us to determine a planar or geometric dual digraph D' that can immediately be viewed as a Kirchhoff dual G'? Can this process always be used to find new Kirchhoff graphs? The Diamond suggests that such a process could be difficult.

8 Kirchhoff graphs and reaction networks

This work now concludes where it began: reaction networks, particularly chemical and electrochemical reaction networks. Indeed, Kirchhoff graphs are called *Kirchhoff* because, as stated in the Introduction, they satisfy the Kirchhoff current and potential laws. This chapter benefits greatly from the work of Baumgartner [1] (2025) in her MQP (senior thesis) in mathematics and chemistry under the supervision of Prof. Ronald L. Grimm and the author.

The Kirchhoff current law means that at each vertex, the reaction rates r_ℓ (currents) for reaction steps heading into and out of the vertex must sum to zero:

$$\sum_\ell \text{sgn}(\ell) r_\ell = 0 \tag{8.1}$$

where ℓ runs through all the steps incident on a given vertex, and $\text{sgn}(\ell) = +1$ if the ℓth step exits the vertex, and $\text{sgn}(\ell) = -1$ if the ℓth step enters the vertex. Likewise, the Kirchhoff potential law means that around each cycle, circuit, or closed walk, the changes of species associated with reaction steps therein cancel, leaving the species as they originally were at the start of the cycle. If the network is conservative and the species concentrations determine the chemical or electrochemical potentials, then this potential is also the same at the beginning and the end of the cycle. If the network is not conservative, then there will still be no net change in the species around a cycle, but the chemical or electrochemical potentials will not be conserved (see Section 8.3).

As discussed in the Introduction, the fact that the network satisfies the Kirchhoff current law at the vertices means that the vertex cuts contain a basis for the row space of the stoichiometric matrix, and the fact that the network satisfies the Kirchhoff potential law means that there is a cycle basis for the null space of the stoichiometric matrix. Because this row space and null space are orthogonal complements, each Kirchhoff graph[1] is *complete* in the sense that it contains all the possible reaction pathways and represents all the rate balances required and guaranteed by the stoichiometry. The various reaction network examples below demonstrate this completeness.

The rest of this chapter mainly considers several reaction networks and their Kirchhoff graphs, showing how all of the above applies. The first section below, however, deals with a more general result: writing the Kirchhoff potential law in terms of reaction rates or rate constants. One specific example, the hydrogen evolution reaction (HER), was already discussed in the Introduction in Section 1.2. The current chapter considers several additional networks and how Kirchhoff graphs can be used to better understand them.

Given the stoichiometric matrix A of a reaction network, a Kirchhoff graph for that matrix serves as a network diagram. With edge vectors corresponding to reaction steps, the Kirchhoff current law is guaranteed by the requirement that all vertex incidence

---

1 As shown earlier, technically all this applies only to vector 2-connected Kirchhoff graphs.

https://doi.org/10.1515/9783111408576-008

vectors lie in the row space of A. To satisfy a form of the Kirchhoff potential law, the labeled edges must satisfy the same dependencies as the columns of A. By constructing a graph on edges that are precisely the column vectors of A, one is *guaranteed* that all cycle vectors will lie in Null(A). As a result, using vectors as edges plays an important role in constructing Kirchhoff graphs.

8.1 Kirchhoff potential law in terms of reaction rates

To express the Kirchhoff potential law in terms of reaction rates, the key considerations are the law of mass action and the Gibbs free energy. According to the *law of mass action*, for any reaction step s_ℓ of the form

$$\sum_i m_i R_i \rightleftharpoons \sum_j n_j P_j \tag{8.2}$$

at equilibrium,

$$K := \frac{k^+}{k^-} = \frac{\prod_j [P_j]^{n_j}}{\prod_i [R_i]^{m_i}} \tag{8.3}$$

where the reaction step index ℓ is suppressed to avoid double indices. Here R_i and P_j are the reactant and product species of the ℓth reaction step s_ℓ, respectively, m_i and n_j are the reactant and product stoichiometric numbers, respectively, k_+ and k_- are the forward and backward reaction rate constants, respectively, and $K := k_+/k_-$ is the equilibrium constant for the reaction step. The concentrations $[\cdot]$ are in some nondimensional units, typically mole fractions. Control parameters such as temperature should be held constant for a specific value of K.

Example 8.1. As a simple example of (8.2) and (8.3), consider the Heyrovsky step from the HER network:

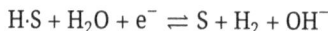

$$\text{H·S} + \text{H}_2\text{O} + \text{e}^- \rightleftharpoons \text{S} + \text{H}_2 + \text{OH}^-$$

By the law of mass action, at equilibrium,

$$K = \frac{[\text{S}][\text{H}_2][\text{OH}^-]}{[\text{H·S}][\text{H}_2\text{O}][\text{e}^-]}$$

where K is some constant. In some cases, some of these concentrations (for example, $[\text{e}^-]$) may be taken to be unity (1) if the species is present in excess/abundance.

According to the Kirchhoff potential law, for a conservative network, the sum of the potential differences ΔV_ℓ around any cycle, circuit, or closed walk must be zero:

$$\sum_\ell \text{sgn}(\ell)\Delta V_\ell = 0 \tag{8.4}$$

where ℓ runs through all the steps in the cycle, circuit, or closed walk, and $\text{sgn}(\ell) = +1$ if the ℓth step is traversed with the same orientation as the edge vector, whereas $\text{sgn}(\ell) = -1$ if the ℓth step is traversed opposite the edge vector orientation. For a chemical reaction network, each of these potential differences ΔV_ℓ is the *Gibbs free energy* ΔG_ℓ for the associated reaction step s_ℓ. The Gibbs free energy can be given *at equilibrium* in terms of the species concentrations:

$$\Delta V_\ell = \Delta G_\ell = RT \ln K_\ell = RT \ln\left(\frac{k_\ell^+}{k_\ell^-}\right) \tag{8.5}$$

where R is the universal gas constant ($8.314\,\text{J}/(\text{mole·K})$), and T is the temperature in Kelvin (K). From (8.4) and (8.5):

$$\sum_\ell \text{sgn}(\ell)\ln\left(\frac{k_\ell^+}{k_\ell^-}\right) = 0$$

which implies

$$\prod_\ell \left(\frac{k_\ell^+}{k_\ell^-}\right)^{\text{sgn}(\ell)} = 1 \tag{8.6}$$

because of the properties of the logarithm function. So the product of the ratios of the forward to the backward rate constants around any closed walk, circuit, or cycle must be unity as long as the Gibbs free energy can be given in terms of the concentrations as in (8.5). Notice that since the species concentrations do not appear in the product (8.6), this rate constant balance must hold even if the reaction network is not in equilibrium.

The rate constant balance condition for cycles in (8.6) can also be stated in terms of reaction rates for the reaction steps. Again by the *law of mass action*, the forward and backward reaction rates are given in terms of the concentrations of the reactants or the products:

$$r^+ = k^+ \prod_i [R_i]^{m_i} \quad \text{and} \quad r^- = k^- \prod_j [P_j]^{n_j}$$

where again the reaction step index ℓ is suppressed to avoid double indices. Since the concentrations of the species must exactly cancel around any closed walk, circuit or cycle, the rate constant balance condition (8.6) can be equally given in terms of the forward and backward reaction rates:

$$\prod_\ell \left(\frac{r_\ell^+}{r_\ell^-}\right)^{\text{sgn}(\ell)} = \prod_\ell \left(\frac{k_\ell^+}{k_\ell^-}\right)^{\text{sgn}(\ell)} = 1 \tag{8.7}$$

Again, the rate balances in (8.7) depend only on the validity of the law of mass action and the Gibbs free energy representation in terms of concentrations, not on the network being in neutral equilibrium (meaning that all reaction steps are in equilibrium). Hence (8.7) can be viewed as a statement of the Kirchhoff potential law. This derivation is not new; see Vilekar et al. [33] (2010) or other treatments of electrochemistry. A specific example involving this statement of the Kirchhoff potential law and (8.7) is given for the hydrogen bromide network in Section 8.4.1.

Finally, notice that the reaction rate for the ℓth reaction step is $r_\ell := r_\ell^+ - r_\ell^-$ where $r_\ell^\pm \geq 0$. Thus the Kirchhoff current law (8.1) can be stated in terms of these forward and backward reaction rates:

$$\sum_\ell \text{sgn}(\ell)r_\ell = 0 \quad \Longrightarrow \quad \sum_\ell \text{sgn}(\ell)r_\ell^+ = \sum_\ell \text{sgn}(\ell)r_\ell^-$$

where again in each sum, ℓ runs through all the steps incident on a given vertex.

8.2 The hydrogen oxidation reaction

The hydrogen oxidation reaction (HOR) was mentioned in the Introduction (Section 1.2) because of its connection to the hydrogen evolution reaction (HER). Both reactions occur in proton exchange membrane fuel cells (PEMFCs) with HER on the alkaline cathode side and HOR on the acidic anode side. The HOR network has three elementary steps leading to one overall reaction:

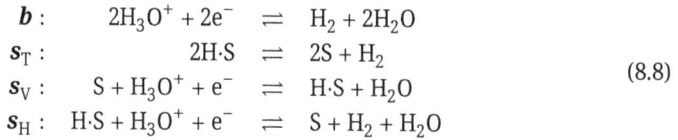

$$\begin{aligned}
b &: & 2H_3O^+ + 2e^- &\rightleftharpoons H_2 + 2H_2O \\
s_T &: & 2H{\cdot}S &\rightleftharpoons 2S + H_2 \\
s_V &: & S + H_3O^+ + e^- &\rightleftharpoons H{\cdot}S + H_2O \\
s_H &: & H{\cdot}S + H_3O^+ + e^- &\rightleftharpoons S + H_2 + H_2O
\end{aligned} \tag{8.8}$$

Here the forward and reverse directions of the steps are written to match those in the HER system from the Introduction. Then the stoichiometric matrix for this HOR network is

$$\begin{array}{c|cccc}
 & b & s_T & s_V & s_H \\
\hline
H_2 & 1 & 1 & 0 & 1 \\
2H_2O & 2 & 0 & 1 & 1 \\
2H_3O^+ & -2 & 0 & -1 & -1 \\
2e^- & -2 & 0 & -1 & -1 \\
S & 0 & 2 & -1 & 1 \\
H{\cdot}S & 0 & -2 & 1 & -1
\end{array} = A$$

where the four reactions (three elementary plus one overall) imply that this A has four columns. This is of course the same stoichiometric matrix obtained for the HER network

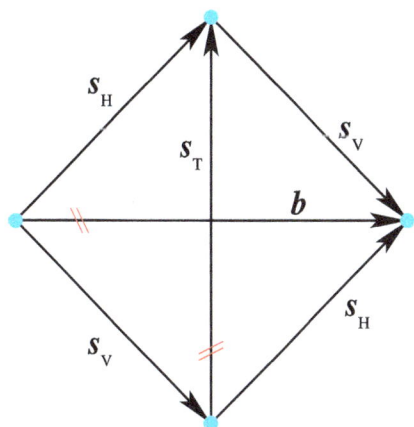

Figure 8.1: A Kirchhoff graph for the HOR reaction network showing how the three elementary steps (vectors) can combine (add) to yield the overall reaction **b**. Again the red hash marks indicate that two copies of s_T and **b** lie in parallel between two pairs of vertices. This is of course the same Kirchhoff graph as the one shown in Figure 1.8.

in (1.5), and so the HOR Kirchhoff graph shown in Figure 8.1 is the same as the HER Kirchhoff graph shown in Figure 1.8, and also the same as the Kirchhoff graph shown in Figure 1.2.

All three the reaction pathways yielding the overall reaction are again present in this graph: (1) $s_H + s_V$, (2) $2s_V + s_T$, and (3) $2s_H - s_T$. The rate balances at the top and bottom vertices again imply that the concentrations of both S and H·S are conserved during the reaction processes. The Wheatstone bridge balance still exists, where the difference in the rates of the Heyrovsky and Volmer steps is given by the Tafel rate. From the right and left vertices we find that the overall reaction rate is the rate of hydrogen production, half the rate of hydronium ion consumption, and the average of the Heyrovsky and Volmer rates:

$$r_{H_2} = -r_{H_3O^+}/2 = r_{OR} = -r_b = (r_V + r_H)/2$$

It is perhaps not surprising that the HER and HOR networks have the same Kirchhoff graphs; these two very similar reaction networks have the same network structure. More surprising may be the structural similarities between the HER/HOR networks and the next reaction network, which at first glance looks very different from these two.

8.3 The Chapman cycle

The Chapman cycle represents the production of ozone in the upper atmosphere (see Dütsch [6] (1968)). It is made up of three elementary steps and one overall reaction:

$$
\begin{aligned}
\boldsymbol{b}: && 3O_2 &\rightleftharpoons 2O_3 \\
\boldsymbol{s}_1: && O_2 &\rightarrow 2O \\
\boldsymbol{s}_2: && O + O_2 &\rightleftharpoons O_3 \\
\boldsymbol{s}_3: && O + O_3 &\rightarrow 2O_2
\end{aligned}
\qquad (8.9)
$$

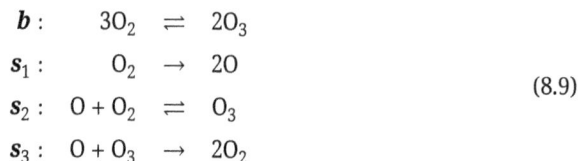

There are three species in this network: molecular oxygen (dioxygen, O_2), ozone (O_3), and free radical oxygen (O). Elementary steps $\boldsymbol{s}_1$ and $\boldsymbol{s}_3$ in this network proceed primarily in the forward direction; the backward rate is essentially zero. The first step $\boldsymbol{s}_1$ is activated by ultraviolet photons with wavelengths below 240 nanometers. The forward direction of step $\boldsymbol{s}_2$ is catalyzed by some other molecule that absorbs excess energy, whereas the backward direction requires another ultraviolet photon and releases heat. Thus this network does not conserve energy, but nonetheless a Kirchhoff graph can be constructed based on the stoichiometry.

The Chapman reaction network (8.9) gives the Chapman stoichiometric matrix:

$$
\begin{array}{c}
\\
O_2 \\
O_3 \\
O
\end{array}
\begin{array}{c}
\begin{array}{cccc}
b & s_1 & s_2 & s_3
\end{array} \\
\left[
\begin{array}{cccc}
-3 & -1 & -1 & 2 \\
2 & 0 & 1 & -1 \\
0 & 2 & -1 & -1
\end{array}
\right] = A
\end{array}
$$

Standard row operations can transform this stoichiometric matrix to the row-equivalent row matrix:

$$
R = \begin{bmatrix} 2 & 0 & 1 & -1 \\ 0 & 2 & -1 & -1 \end{bmatrix}
$$

which of course looks fairly similar to the row matrix for the HER and HOR networks. Indeed the Chapman cycle is K-equivalent to the HER and HOR networks; the third and fourth steps are switched, and the fourth step is reversed. A Kirchhoff graph for the Chapman cycle is shown in Figure 8.2. Bars near the base of steps $\boldsymbol{s}_1$ and $\boldsymbol{s}_3$ indicate that there is no backward direction for these two steps.

Figures 8.1 and 8.2 make clear that these two very different reaction networks have essentially the same reaction structure. The main difference between the two is that two of the steps in the Chapman cycle are irreversible, which changes what reaction pathways exist in the network. The forward overall reaction is achieved only by $\boldsymbol{s}_1 + 2\boldsymbol{s}_2 = \boldsymbol{b}$, whereas the backward overall reaction is achieved either by $\boldsymbol{s}_3 - \boldsymbol{s}_2 = -\boldsymbol{b}$ or by $\boldsymbol{s}_1 + 2\boldsymbol{s}_3 = -\boldsymbol{b}$. Recall that in this network, steps $\boldsymbol{s}_1$ and $\boldsymbol{s}_3$ are irreversible, but step $\boldsymbol{s}_2$ is reversible.

Except for the irreversibility of two of the steps, the Chapman cycle and HER/HOR networks have the same complete structure, meaning that all vertices are adjacent in these Kirchhoff graphs. Still the structure of these networks is rather basic. One might wonder if more complicated networks share similar or identical structures. The next two networks show that this is indeed the case.

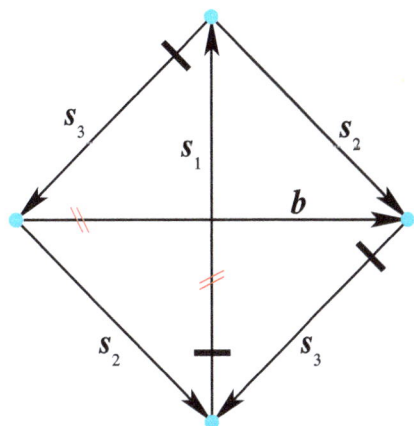

Figure 8.2: A Kirchhoff graph for the Chapman cycle showing how the three elementary steps (vectors) can combine (add) to yield the overall reaction **b**. Because steps s_1 and s_3 are irreversible, there is only one pathway yielding the overall reaction, $b: s_1 + 2s_2 = b$. The black bars near the base of a reaction step indicate that the reverse direction of the reaction step is blocked.

8.4 The hydrogen halide networks

The following set of reaction steps is the hydrogen halide network (HX). In principle, X can be replaced by any of the halogens (halides): fluorine (F), chlorine (Cl), bromine (Br), or iodine (I). Again, the zeroth step **b** is the overall reaction that does not occur by itself, but rather through some sequence of the five elementary steps (s_1–s_5):

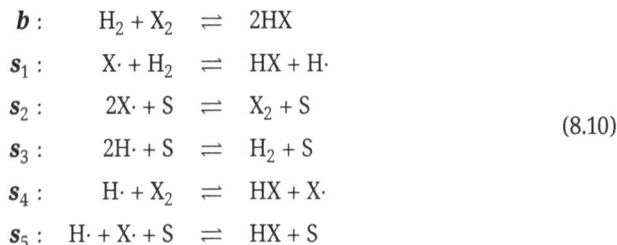

$$
\begin{aligned}
b: \quad & H_2 + X_2 & \rightleftharpoons \quad & 2HX \\
s_1: \quad & X\cdot + H_2 & \rightleftharpoons \quad & HX + H\cdot \\
s_2: \quad & 2X\cdot + S & \rightleftharpoons \quad & X_2 + S \\
s_3: \quad & 2H\cdot + S & \rightleftharpoons \quad & H_2 + S \\
s_4: \quad & H\cdot + X_2 & \rightleftharpoons \quad & HX + X\cdot \\
s_5: \quad & H\cdot + X\cdot + S & \rightleftharpoons \quad & HX + S
\end{aligned}
\tag{8.10}
$$

There are five species in this network: molecular hydrogen (H_2), a molecular halogen (X_2), hydrogen halide (HX), and two radical species: free radical hydrogen (H·) and a free radical halogen (X·). Steps 2, 3, and 5 are catalyzed by S, which does not actually change in any reaction step and so will not be included in the stoichiometry.

The HX stoichiometric matrix in (8.11) can immediately be derived from the HX network (8.10). Its steps and species are labeled as follows:

$$
A = \begin{array}{c} \\ H_2 \\ X_2 \\ HX \\ H \cdot \\ X \cdot \end{array}
\begin{array}{cccccc} b & s_1 & s_2 & s_3 & s_4 & s_5 \\ \end{array}
\left[\begin{array}{cccccc}
-1 & -1 & 0 & 1 & 0 & 0 \\
-1 & 0 & 1 & 0 & -1 & 0 \\
2 & 1 & 0 & 0 & 1 & 1 \\
0 & 1 & 0 & -2 & -1 & -1 \\
0 & -1 & -2 & 0 & 1 & -1
\end{array} \right] = A \tag{8.11}
$$

Elementary row operations can then be used to put this stoichiometric matrix into its reduced echelon form:

$$
R = \begin{bmatrix}
1 & 0 & 0 & 1 & 1 & 1 \\
0 & 1 & 0 & -2 & -1 & -1 \\
0 & 0 & 1 & 1 & 0 & 1
\end{bmatrix} \tag{8.12}
$$

Finally, using the linear programming algorithm from Section 4.2 (or some other appropriate method), one can structure a Kirchhoff graph for this HX network as in Figure 8.3. The overall reaction **b** is doubled in parallel in the middle of the graph, and all the elementary steps appear twice in separate, symmetric locations. Notice that this vector graph is not *complete* in the graphic theoretic sense in that not all vertices are adjacent.

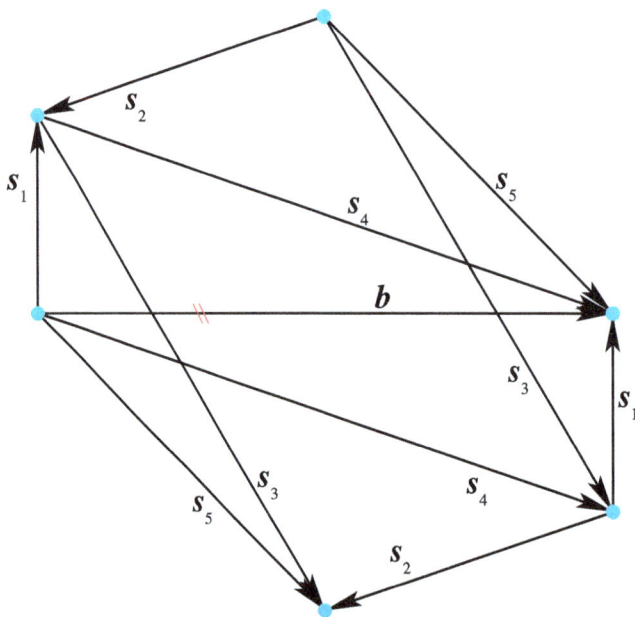

Figure 8.3: A Kirchhoff graph with multiplicity two ($m^* = 2$) for the hydrogen-halide (HX) reaction network. As always, this Kirchhoff graph shows the geometric structure of the HX network.

The reaction pathways for this network can be seen from this Kirchhoff graph in Figure 8.3. Without including any null cycles, the overall reaction b can be achieved by any one of six reaction step combinations:

$$
\begin{aligned}
(1) \quad & b = s_1 + s_4 \\
(2) \quad & b = s_1 - s_2 + s_5 \\
(3) \quad & b = 2s_1 - s_2 + s_3, \\
(4) \quad & b = s_5 - s_3 + s_4 \\
(5) \quad & b = 2s_5 - s_3 - s_2 \\
(6) \quad & b = 2s_4 - s_3 + s_2
\end{aligned}
$$

In principle, this list could be found directly from the HX network (8.10) or its stoichiometric matrix, but it is much more easily seen in the Kirchhoff graph. In practice, one or perhaps two of these routes will dominate, but which ones dominate could change, depending on control parameters such as temperature.

8.4.1 Specifics for the hydrogen bromide (HBr) network

Focusing now on the specific case of the hydrogen bromide network, information about the reaction rates for the HBr network have been measured experimentally by Cooley and Anderson [4] (1952); these values for $T = 1000$ K are shown in Table 8.1. The units for each of these rate constants $k_i^\pm$ are moles/(cm^3 sec). In the Kirchhoff graph in Figure 8.3, there are three cycles involving these five steps:

$$
\begin{aligned}
(1) \quad & 0 = s_1 + s_3 - s_5 \\
(2) \quad & 0 = s_2 + s_4 - s_5 \\
(3) \quad & 0 = s_1 - s_2 + s_3 - s_4
\end{aligned}
$$

The Kirchhoff potential law for rates (8.7) can be applied to these cycles to obtain two equations/constraints that the rate constants much satisfy:

$$
\frac{k_1^+ k_3^+}{k_1^- k_3^-} = \frac{k_5^+}{k_5^-} = \frac{k_2^+ k_4^+}{k_2^- k_4^-} \tag{8.13}
$$

The values given in Table 8.1 satisfy the equation on the left in (8.13) to within the precision of the data, but the equation on the right in (8.13) is *not* satisfied. The values make it somewhat close, but at least one of the rate constants given in Table 8.1 must be incorrect by a factor of about a third if the Kirchhoff potential law holds (8.7).

By the law of mass action, the products of these rate constants with the species concentrations in some convenient nondimensional units (say, mole fractions) yield the forward (+) or backward (−) reaction rates:

Table 8.1: Forward and backward rate constants for each of the five elementary steps of the HBr network as determined by Cooley and Anderson [4].

	k_i^+ [moles/(cm^3 sec)]	k_i^- [moles/(cm^3 sec)]
s_1	$k_1^+ = 9.58 \times 10^{12}$	$k_1^- = 2.70 \times 10^{13}$
s_2	$k_2^+ = 5.7 \times 10^{15}$	$k_2^- = 3.12 \times 10^{13}$
s_3	$k_3^+ = 1.1 \times 10^{16}$	$k_3^- = 8.87 \times 10^{10}$
s_4	$k_4^+ = 2.27 \times 10^{14}$	$k_4^- = 6.18 \times 10^{11}$
s_5	$k_5^+ = 9.0 \times 10^{15}$	$k_5^- = 1.95 \times 10^{11}$

$$
\begin{array}{rclrcl}
r_1^+ &=& k_1^+[\mathrm{Br\cdot}][\mathrm{H_2}], & r_1^- &=& k_1^-[\mathrm{HBr}][\mathrm{H\cdot}], \\
r_2^+ &=& k_2^+[\mathrm{Br\cdot}]^2[\mathrm{S}], & r_2^- &=& k_2^-[\mathrm{Br_2}][\mathrm{S}], \\
r_3^+ &=& k_3^+[\mathrm{H\cdot}]^2[\mathrm{S}], & r_3^- &=& k_3^-[\mathrm{H_2}][\mathrm{S}], \qquad (8.14) \\
r_4^+ &=& k_4^+[\mathrm{Br_2}][\mathrm{H\cdot}], & r_4^- &=& k_4^-[\mathrm{HBr}][\mathrm{Br\cdot}], \\
r_5^+ &=& k_5^+[\mathrm{H\cdot}][\mathrm{Br\cdot}][\mathrm{S}], & r_5^- &=& k_5^-[\mathrm{HBr}][\mathrm{S}].
\end{array}
$$

When the entire network is at neutral equilibrium, ($r_i^+ = r_i^-$ for all $1 \leq i \leq 5$), and these equalities along with (8.14) imply the rate constant constraints (8.13). In addition, neutral equilibrium also yields three additional equalities that constrain the concentrations:

$$
[\mathrm{Br_2}] \;=\; \frac{k_2^+}{k_2^-}[\mathrm{Br\cdot}]^2, \qquad [\mathrm{H_2}] \;=\; \frac{k_3^+}{k_3^-}[\mathrm{H\cdot}]^2, \qquad [\mathrm{HBr}] \;=\; \frac{k_5^+}{k_5^-}[\mathrm{H\cdot}][\mathrm{Br\cdot}] \qquad (8.15)
$$

These of course do depend on concentration, so they only hold at neutral equilibrium.

8.4.2 Steady-state network

Now consider the case where the reaction step reaction rates are not zero: $r_i := r_i^+ - r_i^- \neq 0$ for at least some of the reaction steps (some i). For convenience, assume that these nonzero rates are constant, meaning that the network is at steady state.[2] The rate balances indicated by the vertex cuts at v_2 and v_3 must still hold at steady state. Thus

$$
\begin{array}{rlrcl}
\text{vertex } v_2 : & r_1 + r_2 &=& r_3 + r_4 \\
\text{vertex } v_3 : & -r_2 &=& r_3 + r_5
\end{array} \qquad (8.16)
$$

As in any Kirchhoff graph for a reaction network, the vertex rate balances that do not involve the overall reaction must represent the conservation of the intermediate species (catalysts, radicals, etc.), and so this is the case here. In the reaction network itself (8.10), the rates for the radicals can be given in terms of the reaction step rates:

---

2 This steady state is sometimes referred to as in equilibrium because the time derivatives $dr_i/dt = 0$, but this terminology will be avoided here.

$$r_{X\cdot} = -r_1 - 2r_2 + r_4 - r_5$$
$$r_{H\cdot} = r_1 - 2r_3 - r_4 - r_5 \tag{8.17}$$

A bit of algebra shows that the vertex rate balances in (8.16) imply that $r_{X\cdot} = r_{H\cdot} = 0$ in (8.17).

While the Kirchhoff potential law provides constraints that the rate constants $k_i^{\pm}$ must satisfy, the two Kirchhoff current law constraints in (8.16) define a three-dimensional linear space for $(r_1, r_2, r_3, r_4, r_5)$ as a subspace of $\mathbb{R}^5$. For a steady-state reaction to occur without creating or consuming either of the radicals, the rates $(r_1, r_2, r_3, r_4, r_5)$ must lie in this subspace. On the other hand, each point in this subspace represents a steady-state reaction where the radical and catalyst concentrations remain constant. Now consider the rate balance (vertex cut) condition at v_1:

$$\text{vertex } v_1 : \quad 2r_b = -r_1 - r_4 - r_5 \tag{8.18}$$

If $r_b = 0$, then inside the three-dimensional subspace where the radicals and catalyst are conserved, (8.18) defines a plane. Moreover, $r_b < 0$ in the half-space on one side of this plane, whereas $r_b > 0$ on the other side of the plane. As before, $r_{OR} = -r_b$, so for this reaction network, $r_{HBr} = 2r_{OR} = (r_1 + r_4 + r_5)$. In the half-space where $r_{OR} > 0$, which of the six reaction pathways is dominant in the Kirchhoff graph depends on which specific point in the half-space corresponds to the constant reaction rates.

8.4.3 Additional reaction steps

Suppose that in experimental measurements, one or both conditions in (8.16) are not satisfied. For example, suppose that the rate balance at vertex v_2 is not satisfied:

$$r_1 + r_2 \neq r_3 + r_4$$

Although there could be a number of explanations for this imbalance, one possibility is that the original list of reaction steps (8.10) is missing a step. Recall that the Kirchhoff graph in Figure 8.3 is not *complete* in the graph theory sense since there are vertices that are not adjacent to each other, that is, vertices are not connected by a vector. Suppose that we add a vector connecting vertices v_2 and v_5. Denote this vector by s_+ as is shown in Figure 8.4. This addition has a number of important implications: it increases the number of reaction pathways, for example, adding $2s_4 - s_+$ and $2s_1 + s_+$. Moreover, the rate balance at vertex v_2 becomes:

$$r_1 + r_2 = r_3 + r_4 + r_+$$

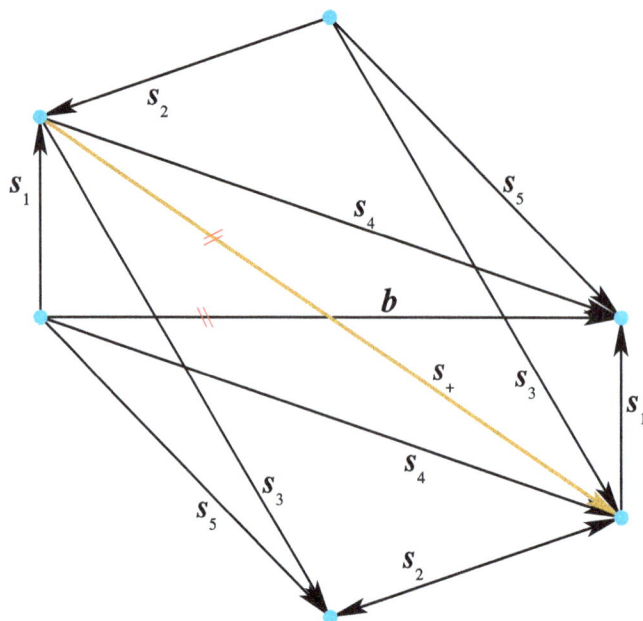

Figure 8.4: A Kirchhoff graph for the extended HBr reaction network. Two copies of the extra reaction step s_+ appear in parallel in the middle of the graph, making this a Kirchhoff graph.

Thus the rate of the new step s_+ might be able to make up the difference between $r_1 + r_2$ and $r_3 + r_4$. Based on the stoichiometry of this reaction network, this new vector s_+ corresponds to the reaction step

$$X_2 + 2H\cdot \rightleftharpoons H_2 + 2X\cdot.$$

This step is somewhat similar to the Heyrovsky step from HER, where three species must come together for the backward reaction, but of course one must determine whether this reaction step is conceivable as an elementary step in this particular network. Still in general, it can be important to know how adding or subtracting a reaction step will affect a network, and a Kirchhoff graph may help clarify this.

8.5 The Langmuir–Hinshelwood network

The basic version of the Langmuir–Hinshelwood network (bLHN) contains only three elementary steps and one overall reaction:

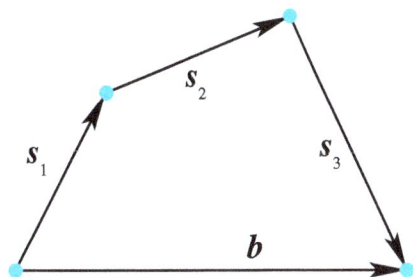

Figure 8.5: Single-cycle Kirchhoff graph for the basic Langmuir-Hinshelwood network.

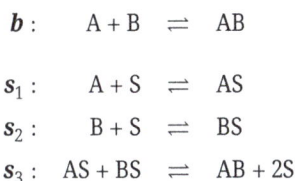

$$b: \qquad A + B \rightleftharpoons AB$$

$$s_1: \qquad A + S \rightleftharpoons AS$$
$$s_2: \qquad B + S \rightleftharpoons BS$$
$$s_3: \quad AS + BS \rightleftharpoons AB + 2S$$

The Kirchhoff graph for this network is a single cycle shown in Figure 8.5. This network has all the advantages and disadvantages associated with a single cycle: easy to understand, but not robust since it relies on a single pathway. All the steps proceed at the same rate.

A much more interesting version of the Langmuir–Hinshelwood network (LHN) occurs when the process requires a reversible activation that can occur anywhere in the network. This version (LHN) is made up of eight elementary steps, one overall reaction and ten species:

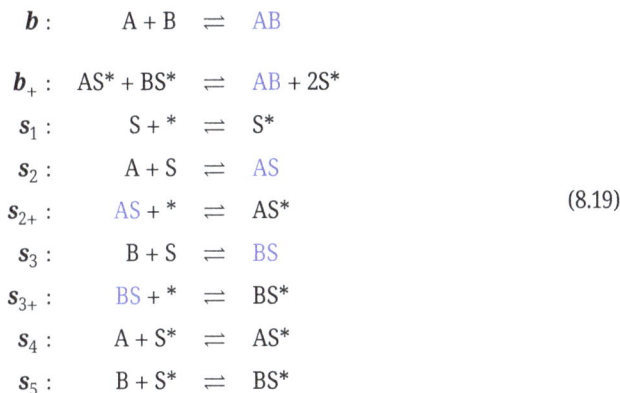

$$
\begin{aligned}
b: & \qquad A + B &\rightleftharpoons& \quad AB \\
b_+: & \quad AS^* + BS^* &\rightleftharpoons& \quad AB + 2S^* \\
s_1: & \qquad S + * &\rightleftharpoons& \quad S^* \\
s_2: & \qquad A + S &\rightleftharpoons& \quad AS \\
s_{2+}: & \qquad AS + * &\rightleftharpoons& \quad AS^* \\
s_3: & \qquad B + S &\rightleftharpoons& \quad BS \\
s_{3+}: & \qquad BS + * &\rightleftharpoons& \quad BS^* \\
s_4: & \qquad A + S^* &\rightleftharpoons& \quad AS^* \\
s_5: & \qquad B + S^* &\rightleftharpoons& \quad BS^*
\end{aligned}
\tag{8.19}
$$

The star * by itself in a reaction step indicates activation, whereas a species with this star indicates the activated version of the species.

On its face, the eight elementary steps of LHN seem to make a Kirchhoff graph for this network much more difficult to construct. Based on the LH network (8.19), its stoichiometric matrix is given in (8.20), where its steps and species are again labeled:

$$
\begin{array}{c}
\quad\quad\;\; b \;\;\; b_+ \;\; s_1 \;\; s_2 \;\; s_{2+} \;\; s_3 \;\; s_{3+} \;\; s_4 \;\; s_5 \\
\begin{array}{c} A \\ B \\ AB \\ S \\ * \\ S^* \\ AS \\ BS \\ AS^* \\ BS^* \end{array}
\left[
\begin{array}{ccccccccc}
-1 & 0 & 0 & -1 & 0 & 0 & 0 & -1 & 0 \\
-1 & 0 & 0 & 0 & 0 & -1 & 0 & 0 & -1 \\
1 & 1 & 0 & 0 & 0 & 0 & 0 & 0 & 0 \\
0 & 0 & -1 & -1 & 0 & -1 & 0 & 0 & 0 \\
0 & 0 & -1 & 0 & -1 & 0 & -1 & 0 & 0 \\
0 & 2 & 1 & 0 & 0 & 0 & 0 & -1 & -1 \\
0 & 0 & 0 & 1 & -1 & 0 & 0 & 0 & 0 \\
0 & 0 & 0 & 0 & 0 & 1 & -1 & 0 & 0 \\
0 & -1 & 0 & 0 & 1 & 0 & 0 & 1 & 0 \\
0 & -1 & 0 & 0 & 0 & 0 & 1 & 0 & 1
\end{array}
\right] = A
\end{array}
\qquad (8.20)
$$

This stoichiometric matrix (8.20) can be row-reduced to yield the row matrix

$$
R = \left[
\begin{array}{ccccccccc}
1 & 0 & 0 & 0 & 0 & 0 & 1 & 0 & 1 \\
0 & 1 & 0 & 0 & 0 & 0 & -1 & 0 & -1 \\
0 & 0 & 1 & 0 & 0 & 0 & 2 & -1 & 1 \\
0 & 0 & 0 & 1 & 0 & 0 & -1 & 1 & -1 \\
0 & 0 & 0 & 0 & 1 & 0 & -1 & 1 & -1 \\
0 & 0 & 0 & 0 & 0 & 1 & -1 & 0 & 0
\end{array}
\right]
\qquad (8.21)
$$

We could attempt to find the Kirchhoff graph directly from R in (8.21), which might be difficult, but fortunately in this case, there are some significant reductions that help make Kirchhoff graph construction easier. Notice that the three species in blue in (8.19) each occur in exactly two steps. This fact allows us for purposes of the construction to combine b with b_+, s_2 with s_{2+}, and s_3 with s_{3+}, thereby eliminating three species (AS, BS, and AB) and three reaction steps from the system. The result is a reduced network, the reduced Langmuir–Hinshelwood network (rLHN):

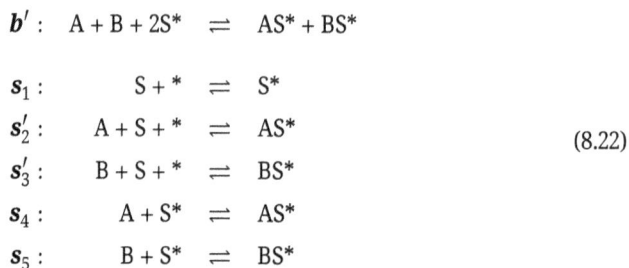

$$
\begin{array}{rl}
b' : & A + B + 2S^* \;\rightleftharpoons\; AS^* + BS^* \\[4pt]
s_1 : & S + * \;\rightleftharpoons\; S^* \\
s_2' : & A + S + * \;\rightleftharpoons\; AS^* \\
s_3' : & B + S + * \;\rightleftharpoons\; BS^* \\
s_4 : & A + S^* \;\rightleftharpoons\; AS^* \\
s_5 : & B + S^* \;\rightleftharpoons\; BS^*
\end{array}
\qquad (8.22)
$$

This reduced network (8.22) has five elementary steps, still one overall reaction and seven species. Then the stoichiometric matrix for this reduced network is

$$
\begin{array}{c c}
 & \begin{array}{cccccc} b' & s_1 & s_2' & s_3' & s_4 & s_5 \end{array} \\
\begin{array}{c} A \\ B \\ S \\ * \\ S^* \\ AS^* \\ BS^* \end{array} &
\left[
\begin{array}{cccccc}
-1 & 0 & -1 & 0 & -1 & 0 \\
-1 & 0 & 0 & -1 & 0 & -1 \\
0 & -1 & -1 & -1 & 0 & 0 \\
0 & -1 & 1 & 1 & 0 & 0 \\
-2 & 1 & 0 & 0 & -1 & -1 \\
1 & 0 & 1 & 0 & 1 & 0 \\
1 & 0 & 0 & 1 & 0 & 1
\end{array}
\right] = A
\end{array}
$$

(8.23)

Finally, row operations again yield the row matrix

$$
R =
\begin{bmatrix}
1 & 0 & 0 & 1 & 0 & 1 \\
0 & 1 & 0 & 2 & -1 & 1 \\
0 & 0 & 1 & -1 & 1 & -1
\end{bmatrix}
$$

(8.24)

The row matrix in (8.24) may look rather different from the row matrix for the hydrogen halide in (8.12), but in fact they are *K-equivalent*. Exchanging/switching the first and third columns of R in (8.24), then multiplying the first, second, and fifth columns by -1 results in a matrix that is row-equivalent to R in (8.12). Hence one Kirchhoff graph for this reduced Langmuir–Hinshelwood network is just a relabeling of one of the fundamental Kirchhoff graphs similar to one for the hydrogen halide network shown in Figure 8.3. This Kirchhoff graph for the reduced network is given in Figure 8.6.

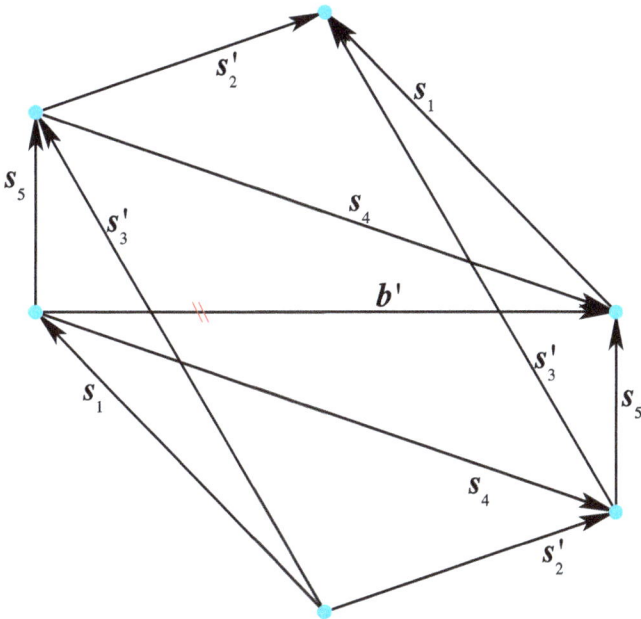

Figure 8.6: A Kirchhoff graph for the reduced Langmuir–Hinshelwood network.

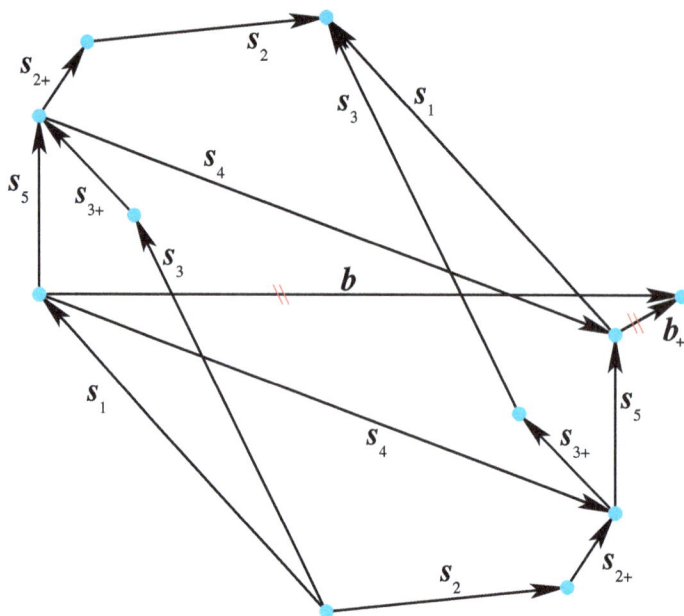

Figure 8.7: A Kirchhoff graph for the full activated Langmuir-Hinshelwood network. This is one of many possible two-dimensional projections of a six-dimensional structure determined by (8.21).

The similarities between the Kirchhoff graphs in Figures 8.3 and 8.6 indicate that these two networks have the same network structure, just as the HER/HOR network and Chapman cycle have the same structure. Now replacing steps b', s_2', and s_3' by the original six vectors, we can reach a Kirchhoff graph for the full activated Langmuir–Hinshelwood network; such a Kirchhoff graph is shown in Figure 8.7.

The Kirchhoff graph for the Langmuir–Hinshelwood network again shows all the possible reaction pathways for this network:

$$
\begin{aligned}
(1)\quad & b = s_5 + s_4 + b_+ \\
(2)\quad & b = s_5 + s_2 + s_{2+} - s_1 + b_+ \\
(3)\quad & b = 2s_5 + s_2 + s_{2+} - s_3 - s_{3+} + b_+ \\
(4)\quad & b = s_4 + s_3 + s_{3+} - s_1 + b_+ \\
(5)\quad & b = 2s_4 + s_3 + s_{3+} - s_2 - s_{2+} + b_+ \\
(6)\quad & b = s_3 + s_{3+} + s_2 + s_{2+} - 2s_1 + b_+
\end{aligned}
$$

Again, which of these pathways is dominant depends on a number of controls such as temperature, but the stoichiometry requires that it appears in this Kirchhoff graph. As was done for the HBr network above, we can also work out all the rate balance requirements for this network set by the Kirchhoff potential and current laws. Because R in

(8.21) has six rows, this Kirchhoff graph naturally embeds in $\mathbb{R}^6$, meaning that there are many projections onto the plane. Figure 8.7 shows just one of many possible choices.

8.6 The NaCl–NaOH network

The final reaction network to be considered is a NaCl–NaOH network that describes the electrolysis of sodium hydroxide from a brine (salt) solution. There are several processes to accomplish this electrolysis (Castner–Kellner, diaphragm, membrane) and several variations of these processes [17], so the steps presented here do not all necessarily occur in all processes, but collecting all these steps together allows us to compare various processes. Our NaCl–NaOH network has eight elementary reaction steps and two overall reactions. This network is the largest we will consider.

The reaction steps and overall reactions for the NaCl–NaOH network are as follows:

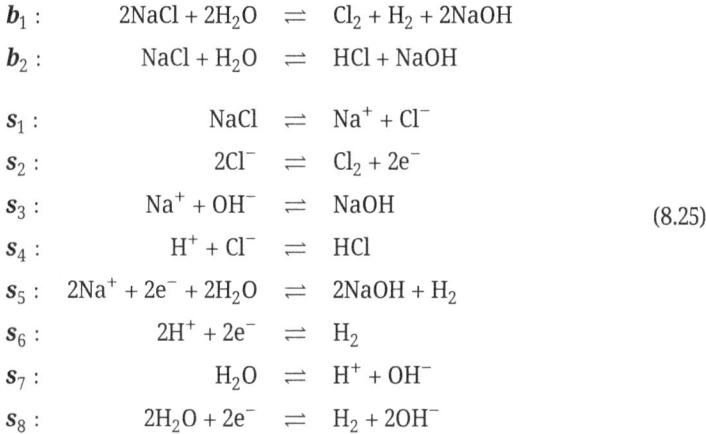

$$
\begin{aligned}
\boldsymbol{b_1} : \quad & 2\mathrm{NaCl} + 2\mathrm{H_2O} \rightleftharpoons \mathrm{Cl_2} + \mathrm{H_2} + 2\mathrm{NaOH} \\
\boldsymbol{b_2} : \quad & \mathrm{NaCl} + \mathrm{H_2O} \rightleftharpoons \mathrm{HCl} + \mathrm{NaOH}
\end{aligned}
$$

$$
\begin{aligned}
\boldsymbol{s_1} : \quad & \mathrm{NaCl} \rightleftharpoons \mathrm{Na^+} + \mathrm{Cl^-} \\
\boldsymbol{s_2} : \quad & 2\mathrm{Cl^-} \rightleftharpoons \mathrm{Cl_2} + 2\mathrm{e^-} \\
\boldsymbol{s_3} : \quad & \mathrm{Na^+} + \mathrm{OH^-} \rightleftharpoons \mathrm{NaOH} \\
\boldsymbol{s_4} : \quad & \mathrm{H^+} + \mathrm{Cl^-} \rightleftharpoons \mathrm{HCl} \\
\boldsymbol{s_5} : \quad & 2\mathrm{Na^+} + 2\mathrm{e^-} + 2\mathrm{H_2O} \rightleftharpoons 2\mathrm{NaOH} + \mathrm{H_2} \\
\boldsymbol{s_6} : \quad & 2\mathrm{H^+} + 2\mathrm{e^-} \rightleftharpoons \mathrm{H_2} \\
\boldsymbol{s_7} : \quad & \mathrm{H_2O} \rightleftharpoons \mathrm{H^+} + \mathrm{OH^-} \\
\boldsymbol{s_8} : \quad & 2\mathrm{H_2O} + 2\mathrm{e^-} \rightleftharpoons \mathrm{H_2} + 2\mathrm{OH^-}
\end{aligned}
\tag{8.25}
$$

These reaction steps are shown schematically in Figure 8.8. The stoichiometric matrix for this network is

	b_1	b_2	s_1	s_2	s_3	s_4	s_5	s_6	s_7	s_8
NaOH	2	1	0	0	1	0	2	0	0	0
NaCl	-2	-1	-1	0	0	0	0	0	0	0
H_2O	-2	-1	0	0	0	0	-2	0	-1	-2
H_2	1	0	0	0	0	0	1	1	0	1
Cl_2	1	0	0	1	0	0	0	0	0	0
HCl	0	1	0	0	0	1	0	0	0	0
H^+	0	0	0	0	0	-1	0	-2	1	0
Na^+	0	0	1	0	-1	0	-2	0	0	0
Cl^-	0	0	1	-2	0	-1	0	0	0	0
OH^-	0	0	0	0	-1	0	0	0	1	2
e^-	0	0	0	2	0	0	-2	-2	0	-2

$$ = A \tag{8.26} $$

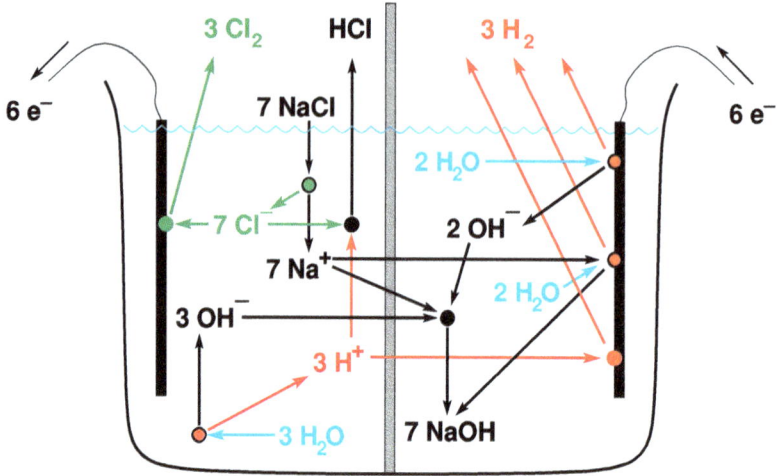

Figure 8.8: A schematic diagram of the eight reaction steps for the NaCl–NaOH network. Each of the steps occurs at one of the dots. Chlorine/chloride is in green; water is in blue; hydrogen is in red. Electrodes are on each side of the container; the anode is on the left, and the cathode is on the right. The grey separator/barrier in the middle keeps hydrogen and chlorine apart. This diagram indicates the net reaction processes, but although all reaction steps involving e^- occur only on the surface of an electrode, other reaction steps occur throughout the electrolyte (water).

and through a sequence of row operations, one can obtain the equivalent row matrix:

$$R = \begin{bmatrix} 1 & 0 & 0 & 0 & 0 & 0 & 1 & 1 & 0 & 1 \\ 0 & 1 & 0 & 0 & 0 & 0 & 0 & -2 & 1 & 0 \\ 0 & 0 & 1 & 0 & 0 & 0 & -2 & 0 & -1 & -2 \\ 0 & 0 & 0 & 1 & 0 & 0 & -1 & -1 & 0 & -1 \\ 0 & 0 & 0 & 0 & 1 & 0 & 0 & 0 & -1 & -2 \\ 0 & 0 & 0 & 0 & 0 & 1 & 0 & 2 & -1 & 0 \end{bmatrix} \tag{8.27}$$

This row matrix (8.27) indicates that $n = 10$ and $k = 6$, that the natural embedding space for this Kirchhoff graph is $\mathbb{R}^6$, and that there are four linearly independent cycles in this graph. Moreover, from this row matrix we can see that there is a basis for the cut space of vertices of degree two and three. This makes it possible (just) to construct a Kirchhoff graph by hand for this network. One such Kirchhoff graph is shown in Figure 8.9. The construction of the Kirchhoff graph in Figure 8.9 begins with drawing the *kernel* in the middle of the graph made up of edge vectors s_4, s_5, s_6, s_7, and s_8. Reaction steps s_1, s_2 and s_3 then link this kernel to the overall reactions b_1 and b_2.

From the Kirchhoff graph in Figure 8.9 we can see that there are only three reaction pathways to achieve overall reaction b_1:

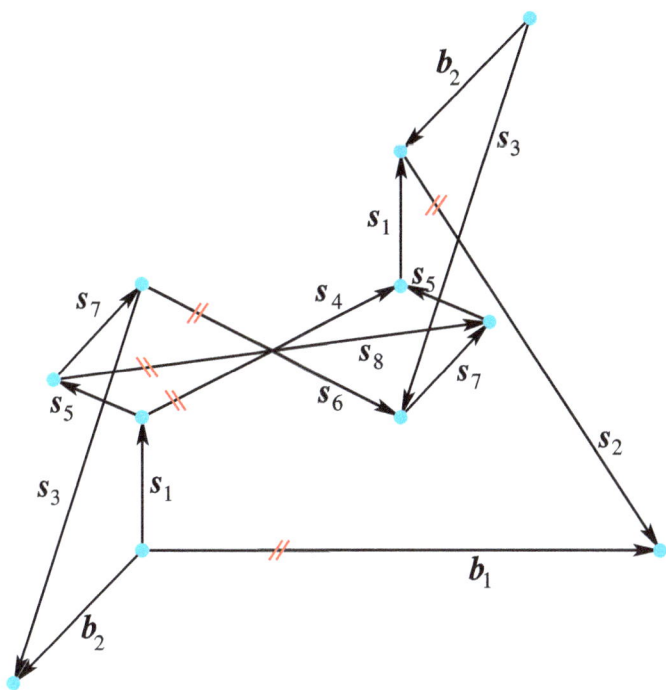

Figure 8.9: A Kirchhoff graph for the NaCl–NaOH network. This graph has multiplicity two ($m^* = 2$). The two overall reactions are shown in the "(x,y)-plane" at the bottom of the graph, but the second copy of b_2 must occur elsewhere in the graph, in this case, at the very top. There is a kernel of reaction steps in the middle of the graph; this kernel is made up of s_4, s_5, s_6, s_7, and s_8. Reaction steps s_1, s_2, and s_3 connect this kernel to the overall reactions.

$$
\begin{aligned}
(1) \quad & b_1 = 2s_1 + s_2 + \ s_4 \\
(2) \quad & b_1 = 2s_1 + s_2 + 2s_5 + s_8 \\
(3) \quad & b_1 = 2s_1 + s_2 + 2s_5 + s_6 + 2s_7
\end{aligned}
\qquad (8.28)
$$

The pathways for b_1 shown in (8.28) do not include the alternate overall reaction b_2 and therefore also do not include s_3. Indeed, if b_2 and the creation of HCl are undesirable, then we must block step s_3. Although this information is in both the list of reaction steps (8.25) and the stoichiometric matrix (8.26), it is difficult or impossible to see there. The Kirchhoff graph makes it clear. On the other hand, blocking step s_3 may be difficult or impossible to accomplish, depending on how control parameters affect the other steps.

Bibliography

[1] G. Baumgartner, *A chemist's guide to Kirchhoff graphs*, MQP (Bachelor's thesis), Worcester Polytechnic Institute (WPI), Worcester, MA, May 2025.

[2] B. Bollobás, *Modern graph theory*, Graduate Texts in Mathematics, vol. 184, Springer-Verlag, New York, NY, 1998, MR 1633290.

[3] A. Cayley, *Desiderata and suggestions: No. 2. The theory of groups: graphical representation*, American Journal of Mathematics **1** (1878), 174–176, MR 1505159, https://doi.org/10.2307/2369306.

[4] S. D. Cooley and R. C. Anderson, *Flame propagation studies using the hydrogen-bromine reaction*, Industrial & Engineering Chemistry **44** (1952), no. 6, 1402–1406, https://doi.org/10.1021/ie50510a058.

[5] R. Diestel, *Graph theory*, Graduate Texts in Mathematics, vol. 173, Springer-Verlag, New York, NY, 1997, MR 1448665.

[6] H. U. Dütsch, *The photochemistry of stratospheric ozone*, Quarterly Journal of the Royal Meteorological Society **94** (1968), no. 402, 483–497, https://doi.org/10.1002/qj.49709440205.

[7] J. D. Fehribach, *Vector-space methods and Kirchhoff graphs for reaction networks*, SIAM Journal Applied Mathematics **70** (2009), 543–562, MR 2524904, https://doi.org/10.1137/080720115.

[8] J. D. Fehribach, *Matrices and their Kirchhoff graphs*, Ars Mathematica Contemporanea **9** (2015), 125–144, MR 3377097, https://doi.org/10.26493/1855-3974.329.82d.

[9] J. D. Fehribach, *Kirchhoff graph uniformity*, Congressus Numerantium **233** (2019), 143–150, MR 4818039, https://combinatorialpress.com/article/cn/Volume%20233/vol-233-paper%2013.pdf.

[10] J. D. Fehribach, *Kirchhoff graphs and stoichiometry*, Journal of The Electrochemical Society **170** (2023), 083504, https://doi.org/10.1149/1945-7111/aceca7.

[11] J. D. Fehribach and J. J. McDonald, *Matrices and Kirchhoff graphs, a rank-two, nullity-two construction*, Congressus Numerantium **230** (2018), 199–207, MR 3967472.

[12] M. C. Gietzmann-Sanders, *Constructing Kirchhoff graphs*, MQP (Bachelor's thesis), Worcester Polytechnic Institute (WPI), Worcester, MA, January 2017.

[13] C. D. Godsil, *Algebraic combinatorics*, Chapman and Hall Mathematics Series, Chapman & Hall, New York, NY, 1993, MR 1220704.

[14] C. D. Godsil and B. D. McKay, *Feasibility conditions for the existence of walk-regular graphs*, Linear Algebra and its Applications **30** (1980), 51–61, MR 568777, https://doi.org/10.1016/0024-3795(80)90180-9.

[15] F. Harary and R. Z. Norman, *Some properties of line digraphs*, Rendiconti del Circolo Matematico di Palermo. Serie II **9** (1960), 161–168, MR 0130839, https://doi.org/10.1007/BF02854581.

[16] W. Magnus, A. Karrass, and D. Solitar, *Combinatorial Group Theory: Presentations of Groups in Terms of Generators And Relations*, 2nd ed., Dover Publications, Inc., New York, NY, 2004, MR 2109550.

[17] C. L. Mantell, *Electrochemical Engineering*, 4th ed., Chemical Engineering Series, ch. 11, McGraw-Hill, New York, NY, 1960.

[18] J. C. Maxwell, *On reciprocal figures, frames, and diagrams of forces*, Transactions of the Royal Society of Edinburgh **26** (1870), 1–40, https://doi.org/10.1017/S0080456800026351.

[19] J. Oxley, *Matroid theory*, 2nd ed., Oxford Graduate Texts in Mathematics, vol. 21, Oxford University Press, Oxford, UK, 2011, MR 2849819.

[20] T. Pisanski and B. Servatius, *Configurations from a graphical viewpoint*, Birkhäuser Advanced Texts: Basler Lehrbücher, Birkhäuser/Springer, New York, NY, 2013, MR 2978043.

[21] L. S. Pitsoulis, *Topics in matroid theory*, SpringerBriefs in Optimization, Springer, New York, NY, 2014, MR 3154793.

[22] A. Recski, *Matroid theory and its applications in electric network theory and in statics*, Algorithms and Combinatorics, vol. 6, Springer-Verlag [Akadémiai Kiadó], Berlin, DE, 1989, MR 1027839, https://doi.org/10.1007/978-3-662-22143-3.

[23] T. Reese, *Kirchhoff graphs*, Ph. D. thesis, Worcester Polytechnic Institute (WPI), Worcester, MA, May 2018, https://web.wpi.edu/Pubs/ETD/Available/etd-032218-131000.

https://doi.org/10.1515/9783111408576-009

[24] T. Reese, J. D. Fehribach, and R. Paffenroth, *Duality in geometric graphs: vector graphs, Kirchhoff graphs and Maxwell reciprocal figures*, Symmetry **8** (2016), 1–28, MR 3475147, https://doi.org/10.3390/sym8030009.

[25] T. Reese, J. D. Fehribach, and R. Paffenroth, *Equitable edge partitions and Kirchhoff graphs*, Linear Algebra and its Applications **639** (2022), 225–242, MR 4368553, https://doi.org/10.1016/j.laa.2022.01.008.

[26] T. Reese, J. D. Fehribach, R. Paffenroth, and B. Servatius, *Matrices over finite fields and their Kirchhoff graphs*, Linear Algebra and its Applications **547** (2018), 128–147, MR 3781365, https://doi.org/10.1016/j.laa.2018.02.020.

[27] T. Reese, J. D. Fehribach, R. Paffenroth, and B. Servatius, *Uniform Kirchhoff graphs*, Linear Algebra and its Applications **566** (2019), 1–16, MR 3894151, https://doi.org/10.1016/j.laa.2018.12.018.

[28] G. Sabidussi, *Vertex-transitive graphs*, Monatshefte für Mathematik **68** (1964), 426–438, MR 0175815, https://doi.org/10.1007/BF01304186.

[29] G. Strang, *Introduction to linear algebra*, 5th ed., Wellesey Cambridge Press, Wellesey, MA, 2016, MR 4886188.

[30] W. T. Tutte, *Matroids and graphs*, Transactions of the American Mathematical Society **90** (1959), 527–552, MR 0101527, https://doi.org/10.2307/1993185.

[31] W. T. Tutte, *An algorithm for determining whether a given binary matroid is graphic*, Proceedings of the American Mathematical Society **11** (1960), 905–917, MR 0117173, https://doi.org/10.2307/2034435.

[32] W. T. Tutte, *Lectures on matroids*, Journal of Research of the National Bureau of Standards **69B** (1965), 1–47, MR 0179781, https://ia600204.us.archive.org/32/items/jresv69Bn1-2p1/jresv69Bn1-2p1_A1b.pdf.

[33] S. A. Vilekar, I. Fishtik, and R. Datta, *Kinetics of the hydrogen electrode reaction*, Journal of The Electrochemical Society **157** (2010), B1040–B1050, https://iopscience.iop.org/article/10.1149/1.3385391/pdf.

[34] J. Wang, *Tiling of prime and composite Kirchhoff graphs*, MQP (Bachelor's thesis), Worcester Polytechnic Institute (WPI), Worcester, MA, March 2023.

[35] J. Wang and J. D. Fehribach, *Prime, composite and fundamental Kirchhoff graphs*, Springer Proceedings in Mathematics & Statistics **462** (2024), 139–148, MR 4841318, https://doi.org/10.1007/978-3-031-62166-6_10.

[36] D. J. A. Welsh, *Matroid theory*, L. M. S. Monographs, vol. 8, Academic Press [Harcourt Brace Jovanovich], New York, NY, 1976, MR 0427112.

[37] H. Whitney, *2-isomorphic graphs*, American Journal of Mathematics **55** (1933), 245–254, MR 1506961, https://doi.org/10.2307/2371127.

Index

https://doi.org/10.1515/9783111408576-010

www.ingramcontent.com/pod-product-compliance
Lightning Source LLC
Chambersburg PA
CBHW081517190326
41458CB00015B/5395